江西财经大学东亿学术论丛·第一辑

稳健混合模型

余 纯 著

Robust Mixture Modeling

本书得到了国家自然科学基金项目——基于非凸惩罚似然法的混合回归模型稳健性推断（项目编号：11661038）的支持。

经济管理出版社
ECONOMY & MANAGEMENT PUBLISHING HOUSE

图书在版编目（CIP）数据

稳健混合模型/余纯著．—北京：经济管理出版社，2019.11
ISBN 978-7-5096-6102-4

Ⅰ.①稳…　Ⅱ.①余…　Ⅲ.①数理统计—统计分析—研究　Ⅳ.①O212

中国版本图书馆 CIP 数据核字（2019）第 244274 号

组稿编辑：王光艳
责任编辑：李红贤
责任印制：黄章平
责任校对：董杉珊

出版发行：经济管理出版社
（北京市海淀区北蜂窝 8 号中雅大厦 A 座 11 层　100038）
网　　址：www. E-mp. com. cn
电　　话：（010）51915602
印　　刷：北京晨旭印刷厂
经　　销：新华书店
开　　本：720mm×1000mm/16
印　　张：10.5
字　　数：183 千字
版　　次：2019 年 12 月第 1 版　　2019 年 12 月第 1 次印刷
书　　号：ISBN 978-7-5096-6102-4
定　　价：68.00 元

江西财经大学统计学院东亿论丛·第一辑
编委会

总主编：罗世华

编　委：罗良清　陶长琪　曹俊文　刘小瑜　魏和清
　　　　平卫英　刘小惠　徐　斌　杨头平　盛积良

总　序

江西财经大学统计学院源于 1923 年成立的江西省立商业学校会统科。统计学专业是学校传统优势专业，拥有包括学士、硕士（含专硕）、博士和博士后流动站的完整学科平台，数量经济学是我校应用经济学下的一个二级学科，拥有硕士、博士和博士后流动站等学科平台。

江西财经大学统计学科是全国规模较大、发展较快的统计学科之一。1978 年、1985 年统计专业分别取得本科、硕士办学权；1997 年、2001 年、2006 年统计学科连续三次被评为省级重点学科；2002 年统计学专业被评为江西省品牌专业；2006 年统计学硕士点被评为江西省示范性硕士点，是江西省第二批研究生教育创新基地。2011 年，江西财经大学统计学院成为我国首批江西省唯一的统计学一级学科博士点授予单位；2012 年，学院获批江西省首个统计学博士后流动站。2017 年，统计学科成功入选“江西省一流学科（成长学科）”；在教育部第四轮学科评估中被评为“A-”等级，进入全国前 10%行列。目前，统计学科是江西省高校统计学科联盟盟主单位，已形成研究生教育为先导、本科教育为主体、国际化合作办学为补充的发展格局。

我们推出这套系列丛书的目的，就是想展现江西财经大学统计学院发展的突出成果，呈现统计学科的前沿理论和方法。之所以以“东亿”冠名，主要是以此感谢高素梅校友及所在的东亿国际传媒给予统计学院的大力支持，在学院发展的关键时期，高素梅校友义无反顾地为我们提供无私的帮助。丛书崇尚学术精神，坚持专业视角，客观务实，兼具科学研究性、实际应用性、参考指导性，希望能给读者以启发和帮助。

丛书的研究成果或结论属个人或研究团队观点，不代表单位或官方结论。如若书中存在不足之处，恳请读者指正。

编委会

2019 年 6 月

Preface

Ordinary Least Squares (OLS) estimators for a linear model are very sensitive to unusual values in the design space or outliers among y values. Even one single atypical value may have a large effect on the parameter estimates. In Chapter 1, we review and describe some available and popular robust techniques, including some recent developed ones, and compare them in terms of breakdown point and efficiency. In addition, we also use simulation studies and real data applications to compare the performance of existing robust methods under different scenarios. Finite mixture models are widely applied in a variety of random phenomena. However, inference of mixture models is a challenging work when the outliers exist in the data. The traditional Maximum Likelihood Estimator (MLE) is sensitive to outliers. In Chapter 2, we give a selective overview of the recently proposed robust mixture regression methods and compare their performance by using simulation studies under different cases. In Chapter 3 and Chapter 4, we propose a Robust Mixture via Mean shift penalization (RMM) in mixture models and Robust Mixture Regression via Mean shift penalization (RM^2) in mixture regression, to achieve simultaneous outlier detection and parameter estimation. A mean shift parameter, which is denoted by γ, is added to the mixture models, and penalized by a nonconvex penalty function. With this model setting, we develop an iterative thresholding embedded EM algorithm to maximize the penalized objective function. Comparing with other existing robust methods, the proposed methods show outstanding performance in both identifying outliers and estimating the parameters.

Contents

Chapter 1
Robust Linear Regression: A Review and Comparison

1.1 Introduction

Linear regression has been one of the most important statistical data analysis tools. Given the independent and identically distributed (*iid*) observations $(\boldsymbol{x}_i, y_i)$, $i=1, \ldots, n$ in order to understand how the response y_is are related to the covariates $\boldsymbol{x}_i$s, we traditionally assume the following linear regression model

$$y_i = \boldsymbol{x}_i^T \boldsymbol{\beta} + \varepsilon_i \tag{1-1}$$

where β is an unknown $p \times 1$ vector, the ε_is are *iid* and independent of $\boldsymbol{x}_i$ with $E(\varepsilon_i | \mathbf{x}_i) = 0$.

The most commonly used estimate for β is the Ordinary Least Square (OLS) estimate which minimizes the sum of squared residuals

$$\sum_{i=1}^{n} (y_i - \boldsymbol{x}_i^T \boldsymbol{\beta})^2 \tag{1-2}$$

However, it is well known that the OLS estimate is extremely sensitive to the outliers. A single outlier can have large effect on the OLS estimate.

In Chapter 1, we review and describe some available robust methods. In addition, a simulation study and a real data application are used to compare different existing robust methods. The efficiency and breakdown point (Donoho and Huber, 1983) are two tradition-ally used important criteria to compare different robust methods. The efficiency is used to measure the relative efficiency of the robust estimate compared to the OLS estimate when the error distribution is exactly normal and there are no outliers. Breakdown point is to measure the proportion of outliers an estimate can tolerate before it goes to infinity. Finite sample breakdown point (Donoho and Huber, 1983) is used and defined as follows: Let $z_i = (\boldsymbol{x}_i, y_i)$. Given any sample $\boldsymbol{z} = (z_i, \ldots, z_n)$, denote $T(\boldsymbol{z})$ the estimate of the parameter $\boldsymbol{\beta}$. Let $\boldsymbol{z}'$ be the corrupted sample where any m of the original points of $\boldsymbol{z}$ are replaced by arbitrary bad data. Then the finite sample breakdown point δ^* is defined as

$$\delta^*(\mathbf{z}, T) = \min_{1 \leqslant m \leqslant n} \left\{ \frac{m}{n} : \sup_{\mathbf{z}'} \| T(\mathbf{z}') - T(\mathbf{z}) \| = \infty \right\} \tag{1-3}$$

where $\|\cdot\|$ is the Euclidean norm.

Many robust methods have been proposed to achieve high breakdown point or high efficiency or both. M-estimates (Huber, 1981) are solutions of the normal equation with appropriate weight functions. They are resistant to unusual y observations, but sensitive to high leverage points on $\boldsymbol{x}$. Hence the breakdown point of an M-estimate is $1/n$. R-estimates (Jackel, 1972) which minimize the sum of scores of the ranked residuals have relatively high efficiency, but their breakdown points are as low as those of OLS estimates. Least Median of Squares (LMS) estimates (Siegel, 1982) which minimize the median of squared residuals, Least Trimmed Squares (LTS) estimates (Rousseeuw, 1983) which minimize the trimmed sum of squared residuals, and S-estimates (Rousseeuw and Yohai, 1984) which minimize the variance of the residuals, all have high breakdown point but with low efficiency. Generalized S-estimates (GS-estimates) (Croux et al., 1994) maintain high breakdown point as S-estimates and have slightly higher efficiency. MM-estimates proposed by Yohai (1987) can simultaneously attain high breakdown point and efficiencies. Mallows Generalized M-estimates (Mallows, 1975) and Schweppe Generalized M-estimates (Handschin et al., 1975) downweight the high leverage points on $\boldsymbol{x}$ but cannot distinguish "good" and "bad" leverage points, thus resulting in a loss of efficiencies. In addition, these two estimators have low breakdown points when p, the number of explanatory variables, is large. Schweppe one-step (S1S) Generalized M-estimates (Coakley and Hettmansperger, 1993) overcome the problems of Schweppe Generalized M-estimates and are calculated in one step. They both have high breakdown points and high efficiencies. Recently, Gervini and Yohai (2002) proposed a new class of high breakdown point and high efficiency robust estimate called robust and efficient weighted least squares estimator (REWLSE). Lee et al. (2012) and She and Owen (2011) proposed a new class of robust methods based on the regularization of case-specific parameters for each response. They further proved that the M-estimator with Huber's ψ function is a special case of their proposed estimator. Wilcox (1996) and You (1999) provided an excellent Monte Carlo comparison of some of the mentioned robust methods. This article aims to provide a more complete review and comparison of existing robust methods including some recently developed robust methods. In addition, besides comparing regression estimates for different robust methods, we also add the comparison of their performance of outlier detection based on three criteria used by She and Owen (2011).

The rest of Chapter 1 is organized as follows. In Section 1.2, we review and describe some of the available robust methods. In Section 1.3, simulation studies and real data applications are used to compare different robust methods. Some discussions are given in Section 1.4.

1.2 Robust Regression Methods

1.2.1 M-estimates

By replacing the least squares criterion (1-2) with a robust criterion, M-estimate (Huber, 1964) of $\boldsymbol{\beta}$ is

$$\hat{\boldsymbol{\beta}} = \arg\min_{\boldsymbol{\beta}} \sum_{i=1}^{n} \rho\left(\frac{y_i - \boldsymbol{x}_i^T\boldsymbol{\beta}}{\hat{\sigma}}\right) \tag{1-4}$$

where $\rho(\cdot)$ is a robust loss function and $\hat{\sigma}$ is an error scale estimate. The derivative of ρ, denoted by $\psi(\cdot)=\rho'(\cdot)$, is called the influence function. In particular, if $\rho(t)=\frac{1}{2}t^2$, then the solution is the OLS estimate.

The OLS estimate is very sensitive to outliers. Rousseeuw and Yohai (1984) indicated that OLS estimates have a breakdown point (BP) of BP = $1/n$ which tends to zero when the sample size n is getting large. Therefore, one single unusual observation can have large impact on the OLS estimate.

One of the commonly used robust loss functions is Huber's ψ function (Huber, 1981), where $\psi_c(t)=\rho'(t)=\max\{-c, \min(c, t)\}$. Huber (1981) recommends using $c=1.345$ in practice. This choice produces a relative efficiency of approximately 95% when the error density is normal. Another possibility for $\psi(\cdot)$ is Tukey's bisquare function $\psi_c(t)=t\{1-(t/c)^2\}_+^2$. The use of $c=4.685$ produces 95% efficiency.

If $\rho(t)=|t|$, then least absolute deviation (LAD, also called median regression) estimates are achieved by minimizing the sum of the absolute values of the residuals

$$\hat{\boldsymbol{\beta}} = \arg\min_{\boldsymbol{\beta}} \sum_{i=1}^{n} |y_i - \boldsymbol{x}_i^T \boldsymbol{\beta}| \tag{1-5}$$

The LAD is also called L_1 estimate due to the L_1 norm used. Although LAD is more resistent than OLS to unusual y values, it is sensitive to high leverage outliers, and thus having a breakdown point of BP $= 1/n \to 0$ (Rousseeuw and Yohai, 1984). Moreover, LAD estimates have a low efficiency of 64% when the errors are normally distributed. Similar to LAD estimates, the general monotone M-estimates, i.e., M-estimates with monotone ψ functions, have a BP $= 1/n \to 0$ due to lack of immunity to high leverage outliers (Maronna et al. 2006).

1.2.2 LMS estimates

The LMS estimates (Siegel, 1982) are found by minimizing the median of the squared residuals

$$\hat{\boldsymbol{\beta}} = \arg\min_{\boldsymbol{\beta}} \operatorname{Med}\{(y_i - \boldsymbol{x}_i^T\boldsymbol{\beta})^2\} \tag{1-6}$$

One good property of the LMS estimate is that it possesses a high breakdown point of near 0.5. However, LMS estimates do not have a well-defined influence function because of its convergence rate of $n^{-\frac{1}{3}}$ and thus have a very low efficiency (Rousseeuw, 1984). You (1999) points out that LMS estimates have a zero efficiency. Despite these limitations, the LMS estimate can be used as an initial estimate for some other high breakdown point and high efficiency robust methods.

1.2.3 LTS estimates

The LTS estimate (Rousseeuw, 1983) is defined as

$$\hat{\boldsymbol{\beta}} = \arg\min_{\boldsymbol{\beta}} \sum_{i=1}^{q} r_{(i)}(\boldsymbol{\beta})^2 \tag{1-7}$$

where $r_{(i)}(\beta) = y_{(i)} - \boldsymbol{x}_{(i)}^T\beta$, $r_{(1)}(\beta)^2 \leqslant \cdots \leqslant r_{(q)}(\beta)^2$ are ordered squared residuals, $q = [n(1-\alpha)+1]$, and α is the proportion of trimming. Using $q = \left(\frac{n}{2}\right)+1$ ensures that the estimator has a breakdown point of BP = 0.5, and the convergence rate of $n^{-\frac{1}{2}}$ (Rousseeuw, 1983). Although highly resistent to outliers, LTS suffers badly in terms of very low efficiency, which is about 0.08, relative to OLS estimates (Stromberg et al. 2000). The reason that LTS estimates call attentions to us is that it is traditionally used as the initial estimate for some other high breakdown point and high efficiency robust methods.

1.2.4 S-estimates

S-estimates (Rousseeuw and Yohai, 1984) are defined by

$$\hat{\boldsymbol{\beta}} = \arg\min_{\boldsymbol{\beta}} \hat{\sigma}\left(r_1\left(\boldsymbol{\beta}\right), \cdots, r_n\left(\boldsymbol{\beta}\right)\right) \tag{1-8}$$

where $r_i(\boldsymbol{\beta}) = y_i - \boldsymbol{x}_i^T \boldsymbol{\beta}$ and $\hat{\sigma}(r_1(\boldsymbol{\beta}), \cdots, r_n(\boldsymbol{\beta}))$ is the scale M-estimate which is defined as the solution of

$$\frac{1}{n}\sum_{i=1}^{n} \rho\left(\frac{r_i\left(\boldsymbol{\beta}\right)}{\hat{\sigma}}\right) = \delta \tag{1-9}$$

for any given $\boldsymbol{\beta}$ where δ is taken to be $E_\Phi[\rho(r)]$. For the biweight scale, S-estimates can attain a high breakdown point of BP = 0.5 and has an asymptotic efficiency of 0.29 under the assumption of normally distributed errors (Maronna et al., 2006).

1.2.5 Generalized S-estimates (GS-estimates)

Croux et al. (1994) proposed generalized S-estimates in an attempt to improve the low efficiency of S-estimators. Generalized S-estimates are defined as

$$\hat{\boldsymbol{\beta}} = \arg\min_{\boldsymbol{\beta}} S_n(\boldsymbol{\beta}) \tag{1-10}$$

where $S_n(\boldsymbol{\beta})$ is defined as

$$S_n(\boldsymbol{\beta}) = \sup\left\{S > 0; \binom{n}{2}^{-1} \sum_{i<j} \rho\left(\frac{r_i - r_j}{S}\right) \geqslant k_{n,p}\right\} \tag{1-11}$$

where $r_i = y_i - \boldsymbol{x}_i^T \boldsymbol{\beta}$ p is the number of regression parameters, and $k_{n,p}$ is a constant which might depend on n and p. Particularly, if $\rho(x) = I(|x| \geqslant 1)$ and $k_{n,p} = \left(\binom{n}{2} - \binom{h_p}{2} + 1\right) \Big/ \binom{n}{2}$ with $h_p = \frac{n+p+1}{2}$, generalized S-estimator yields a special case, the least quartile difference (LQD) estimator, which is defined as

$$\hat{\boldsymbol{\beta}} = \arg\min_{\boldsymbol{\beta}} Q_n(r_1, \ldots, r_n) \tag{1-12}$$

where

$$Q_n = \{|r_i - r_j|; i < j\}_{\binom{h_p}{2}} \tag{1-13}$$

is the $\binom{h_p}{2}$th order statistic among the $\binom{n}{2}$ elements of the set $\{|r_i - r_j|; i < j\}$. Generalized S-estimates have a breakdown point as high as S-estimates and with a higher efficiency.

1.2.6 MM-estimates

First proposed by Yohai (1987), MM-estimates have become increasingly popular and are one of the most commonly employed robust regression techniques. The MM-estimates can be found by a three-stage procedure. In the first stage, compute an initial consistent estimate $\hat{\beta}_0$ with high breakdown point but possibly low normal efficiency. In the second stage, compute a robust M-estimate of scale $\hat{\sigma}$ of the residuals based on the initial estimate. In the third stage, find an M-estimate $\hat{\beta}$ starting at $\hat{\beta}_0$.

In practice, LMS or S-estimate with Huber or bisquare functions is typically used as the initial estimate $\hat{\beta}_0$. Let $\rho_0(r) = \rho_1(r/k_0)$, $\rho(r) = \rho_1(r/k_1)$, and assume that each of the ρ-functions is bounded. The scale estimate $\hat{\sigma}$ satisfies

$$\frac{1}{n}\sum_{i=1}^{n} \rho_0\left(\frac{r_i\left(\hat{\boldsymbol{\beta}}\right)}{\hat{\sigma}}\right) = 0.5 \tag{1-14}$$

If the ρ-function is biweight, then $k_0 = 1.56$ ensures that the estimator has the asymptotic BP = 0.5. Note that an M-estimate minimizes

$$L(\beta)=\sum_{i=1}^{n}\rho\left(\frac{r_i\left(\hat{\beta}\right)}{\hat{\sigma}}\right) \tag{1-15}$$

Let ρ satisfy $\rho\leqslant\rho_0$. Yohai (1987) showed that if $\hat{\beta}$ satisfies $L(\hat{\beta})\leqslant(\hat{\beta}_0)$, then $\hat{\beta}$'s BP is not less than that of $\hat{\beta}_0$. Furthermore, the breakdown point of the MM-estimate depends only on k_0 and the asymptotic variance of the MM-estimate depends only on k_1. We can choose k_1 in order to attain the desired normal efficiency without affecting its breakdown point. In order to let $\rho\leqslant\rho_0$, we must have $k_1\geqslant k_0$; the larger the k_1 is, the higher efficiency the MM-estimate can attain at the normal distribution.

provides the values of k_1 with the corresponding efficiencies of the biweight ρ-function. Please see the following table for more detail.

Efficiency	0.80	0.85	0.90	0.95
k_1	3.14	3.44	3.88	4.68

However, Yohai (1987) indicates that MM-estimates with larger values of k_1 are more sensitive to outliers than the estimates corresponding to smaller values of k_1. In practice, an MM-estimate with bisquare function and efficiency 0.85 ($k_1=3.44$) starting from a bisquare S-estimate is recommended.

1.2.7 Mallows GM-estimates

In order to make M-estimates resistent to high leverage outliers, Mallows (1975) proposed Mallows GM-estimates which are defined by

$$\sum_{i=1}^{n} w_i \psi \left\{ \frac{r_i\left(\hat{\beta}\right)}{\hat{\sigma}} \right\} \boldsymbol{x}_i = 0 \tag{1-16}$$

where $\psi(e)=\rho'(e)$ and $w_i=\sqrt{1-h_i}$ with h_i being the leverage of the i th observation. The weight w_i ensures that the observation with high leverage receives less weight than observation with small leverage. However, even "good" leverage points that fall in line with the pattern in the bulk of the data are down-weighted, resulting in a loss of efficiency.

1.2.8 Schweppe GM-estimates

Schweppe GM-estimate (Handschin et al., 1975) is defined by the solution of

$$\sum_{i=1}^{n} w_i \psi \left\{ \frac{r_i\left(\hat{\beta}\right)}{w_i \hat{\sigma}} \right\} \boldsymbol{x}_i = 0 \tag{1-17}$$

which adjusts the leverage weights according to the size of the residual r_i. Carroll and Welsch (1988) proved that the Schweppe estimator is not consistent when the errors are asymmetric. Furthermore, the breakdown points for both Mallows and Schweppe GM-estimates are no more than $1/(p+1)$, where p is the number of unknown parameters.

1.2.9 S1S GM-estimates

Coakley and Hettmansperger (1993) proposed Schweppe one-step (S1S) estimates, which extend from the original Schweppe estimator. S1S estimator is defined as

$$\hat{\boldsymbol{\beta}} = \hat{\boldsymbol{\beta}}_0 + \left[\sum_{i=1}^{n} \psi' \left(\frac{r_i\left(\hat{\boldsymbol{\beta}}_0\right)}{\hat{\sigma} w_i}\right) \boldsymbol{x}_i \boldsymbol{x}_i'\right]^{-1} \times \sum_{i=1}^{n} \hat{\sigma} w_i \psi \left(\frac{r_i\left(\hat{\boldsymbol{\beta}}_0\right)}{\hat{\sigma} w_i}\right) \boldsymbol{x}_i \tag{1-18}$$

where the weight w_i is defined in the same way as Schweppe's GM-estimate.

The method for S1S estimate is different from the Mallows and Schweppe GM-estimates in that once the initial estimates of the residuals and the scale of the residuals are given, final M-estimates are calculated in one step rather than iteratively. Coakley and Hettmansperger (1993) recommended to use Rousseeuw's LTS for the initial estimates of the residuals and LMS for the initial estimates of the scale, and proved that the S1S estimate gives a breakdown point of BP = 0.5 and results in 0.95 efficiency compared to the OLS estimate under the Gauss-Markov assumption.

1.2.10 R-estimates

The R-estimates (Jackel, 1972) minimize the sum of some scores of the ranked residuals

$$\min \sum_{i=1}^{n} a_n (R_i) r_i \tag{1-19}$$

where R_i represents the rank of the i th residual r_i, and $a_n(\cdot)$ is a monotone score function that satisfies

$$\sum_{i=1}^{n} a_n (i) = 0 \tag{1-20}$$

R-estimates are scale equivalent which is an advantage compared to M-estimates. However, the optimal choice of the score function is unclear. In addition, most of R-estimates have a breakdown point of BP = $1/n \to 0$. The bounded influence R-estimator proposed by Naranjo and Hettmansperger (1994) has a fairly high efficiency when the errors have normal distribution. However, it is proved that their breakdown point is no more than 0.2.

1.2.11 REWLSE

Gervini and Yohai (2002) proposed a new class of robust regression method called robust and efficient weighted least squares estimator (REWLSE). REWLSE is much more attractive than many other robust estimators due to its simultaneously attaining maximum breakdown point and full efficiency under normal errors. This new estimator is a type of weighted least squares estimator with the weights adaptively calculated from an initial robust estimator.

Consider a pair of initial robust estimates of regression parameters and scale, $\hat{\beta}_0$ and $\hat{\sigma}$ respectively, the standardized residuals are defined as

$$r_i = \frac{y_i - \boldsymbol{x}_i^T \hat{\boldsymbol{\beta}}_0}{\hat{\sigma}}$$

A large value of $|r_i|$ would suggest that $(\boldsymbol{x}_i, y_i)$ is an outlier.

Define a measure of proportion of outliers in the sample

$$d_n = \max_{i > i_0} \left\{ F^+(|r|_{(i)}) - \frac{(i-1)}{n} \right\}^+ \tag{1-21}$$

where $\{\cdot\}^+$ denotes positive part, F^+ denotes the distribution of $|X|$ when $X \sim F$, $|r|_{(1)} \leqslant \ldots \leqslant |r|_{(n)}$ are the order statistics of the standardized absolute residuals, and $i_0 = \max\{i : |r|_{(i)} < \eta\}$, where η is some large quantile of F^+. Typically $\eta =$

2.5 and the cdf of a normal distribution is chosen for F. Thus those $\lfloor nd_n \rfloor$ observations with largest standardized absolute residuals are eliminated (here $\lfloor a \rfloor$ is the largest integer less than or equal to a).

The adaptive cut-off value is $t_n = |r|_{(in)}$ with $i_n = n - \lfloor nd_n \rfloor$. With this adaptive cut-off value, the adaptive weights proposed by Gervini and Yohai (2002) are

$$w_i = \begin{cases} 1, & \text{if } |r_i| < t_n \\ 0, & \text{if } |r_i| \geqslant t_n \end{cases} \tag{1-22}$$

Then, the REWLSE is

$$\hat{\beta} = (\mathbf{X}^T\mathbf{W}\mathbf{X})^{-1}\mathbf{X}^T\mathbf{W}\mathbf{y}, \tag{1-23}$$

where $\mathbf{W} = \mathrm{diag}(w_1, \cdots, w_n)$, $\mathbf{X} = (\mathbf{x}_1, \ldots, \mathbf{x}_n)^T$, and $\mathbf{y} = (y_1, \cdots, y_n)'$.

If the initial regression and scale estimates with BP = 0.5 are chosen, the breakdown point of the REWLSE is also 0.5. Furthermore, when the errors are normally distributed, the REWLSE is asymptotically equivalent to the OLS estimates and hence asymptotically efficient.

1.2.12 Robust regression based on regularization of case-specific parameters

She and Owen (2011) and Lee et al. (2012) proposed a new class of robust regression methods using the case-specific indicators in a mean shift model with regularization method. A mean shift model for the linear regression is

$$\mathbf{y} = \boldsymbol{X}\boldsymbol{\beta} + \boldsymbol{\gamma} + \boldsymbol{\varepsilon},\ \boldsymbol{\varepsilon} \sim N(0, \sigma^2 I)$$

where $\mathbf{y} = (y_1, \cdots, y_n)^T$, $\boldsymbol{X} = (\boldsymbol{x}_1, \ldots, \boldsymbol{x}_n)^T$, and $\boldsymbol{\gamma} = (\gamma_1, \ldots, \gamma_n)^T$. The mean shift parameter γ_i is nonzero when the i th observation is an outlier and zero, otherwise.

Due to the sparsity of γ_is, She and Owen (2011) and Lee et al. (2012) proposed to estimate β and γ by minimizing the penalized least squares using L_1 penalty

$$L(\beta,\gamma)=\frac{1}{2}\{\mathbf{y}-(X\beta+\gamma)\}^T\{\mathbf{y}-(X\beta+\gamma)\}+\lambda\sum_{i=1}^{n}|\gamma_i| \tag{1-24}$$

where λ are fixed regularization parameters for γ. Given the estimate $\hat{\gamma}$, $\hat{\beta}$ is the OLS estimate with y replaced by $y-\gamma$. For a fixed $\hat{\beta}$, the minimizer of (1-24) is $\hat{\gamma}_i=sgn(r_i)(|r_i|-\lambda)_+$, that is

$$\hat{\gamma}_i=\begin{cases}0, & \text{if } |r_i|\leqslant\lambda\\ y_i-\boldsymbol{x}_i^T\hat{\beta}, & \text{if } |r_i|>\lambda.\end{cases}$$

Therefore, the solution of (1-24) can be found by iteratively updating the above two steps. She and Owen (2011) and Lee et al. (2012) proved that the above estimate is in fact equivalent to the M-estimate if Huber's ψ function is used. However, their proposed robust estimates are based on different perspective and can be extended to many other likelihood based models.

Note, however, the monotone M-estimate is not resistent to the high leverage outliers. In order to overcome this problem, She and Owen (2011) further proposed to replace the L_1 penalty in (1-24) by a general penalty. The objective function is then defined by

$$L_p(\boldsymbol{\beta},\gamma)=\frac{1}{2}\{\mathbf{y}-(\boldsymbol{X\beta}+\gamma)\}^T\{\mathbf{y}-(\boldsymbol{X\beta}+\gamma)\}+\sum_{i=1}^{n}p_\lambda(|\gamma_i|) \tag{1-25}$$

where $p_\lambda(|\cdot|)$ is any penalty function which depends on the regularization parameter λ. We can find $\hat{\gamma}$ by defining a thresholding function $\Theta(\gamma; \lambda)$ (She, 2009). She (2009) and She and Owen (2011) proved that for a specific thresholding function, we can always find the corresponding penalty function. For example, the soft, hard, and smoothly clipped absolute deviation (SCAD, Fan and Li (2001)) thresholding solutions of γ correspond to L_1, Hard, and SCAD penalty functions, respectively. Minimizing the equation (1-25) yields a sparse $\hat{\gamma}$ for outlier detection and a robust estimate of $\boldsymbol{\beta}$. She and Owen (2011) showed that the proposed estimates of (1-25) with hard or SCAD penalties are equivalent to the M-estimates with certain redescending ψ functions and thus will be resistent to high leverage outliers if a high breakdown point robust estimates are used as the initial values.

1.3 Examples

In this section, we use both simulation study and real data applications to compare different robust methods in terms of parameter estimation and outlier detection. The first statistical criterion we use to compare different estimates is the mean squared errors (MSE). The other two are robust measures: robust bias (RB) and median absolute deviation (MAD) (You, 1999). They are defined as

$$RB_i = \text{median}(\hat{\beta}_i) - \beta_i$$

and

$$MAD_i = \text{median}(|\hat{\beta}_i - \beta_i|)$$

where $i = 0, 1$ for Example 1 and $i = 0, 1, 2, 3$ for Example 2.

We compare the OLS estimate with seven other commonly used robust regression estimates: the M estimate using Huber's ψ function (M_H), the M estimate using Tukey's bisquare function (M_T), the S estimate, the LTS estimate, the LMS estimate, the MM estimate (using bisquare weights and $k_1 = 4.68$), and the REWLSE. Due to the limited space, we cannot include all of our reviewed methods for comparison in our simulation study, such as median regression, Mallows GM-estimate, Schweppe GM-estimate, the S1S-estimator, and R-estimates. We mainly choose some methods that are popularly used and can be found from existing R packages. *R* function *rlm* provides the implementation of M_H and M_T with stating psi function as Huber and Tukey, respectively. LMS, LTS and S are computed using *R* function *lqs* with the option specified as "lms" "lts" and "S", respectively. In these *lqs* computation procedures, resampling algorithm is used. *R* package *robust* provides the implementation of MM and REWLSE, and the S-estimate is used as an initial estimate via random resampling. It is known that using the initial S estimate in two-stage algorithm of MM achieves both high efficiency and robustness (Yohai, 1987). Note that we did not include the case-specific regularization methods proposed by She and Owen (2011) and Lee et al. (2012) since they are essentially equivalent to M-estimators. All the computations in this article are done by R. However, one can also implement those robust regression methods using SAS. In SAS, "ROBUSTREG" procedure provides implementation of M, LTS, S and MM estimates choosing "method=" to be "m","lts","s" or "mm", respectively.

Example 1. 1 We generate n samples $\{(x_1, y_1), \dots, (x_n, y_n)\}$ from the model

$$Y=X+\varepsilon$$

where $X\sim N(0, 1)$. In order to compare the performance of different methods, we consider the following six cases for the error density of ε:

Case Ⅰ: $\varepsilon\sim N(0, 1)$-standard normal distribution.

Case Ⅱ: $\varepsilon\sim t_3$-t-distribution with degrees of freedom 3.

Case Ⅲ: $\varepsilon\sim t_1$-t-distribution with degrees of freedom 1 (Cauchy distribution).

Case Ⅳ: $\varepsilon\sim 0.95N(0, 1)+0.05N(0, 10^2)$ -contaminated normal mixture.

Case Ⅴ: $\varepsilon\sim N(0, 1)$ with 10% identical outliers in y direction (where we let the first 10% of y's equal to 30).

Case Ⅵ: $\varepsilon\sim N(0, 1)$ with 10% identical high leverage outliers (where we let the first 10% of x's equal to 10 and their corresponding y's equal to 50).

Table 1. 1 and Table 1. 2 report MSE, RB and MAD of the parameter estimates for each estimation method with sample size $n=20$ and 100, respectively. The number of replicates is 200. From the tables, we can see that MM and REWLSE have the overall best performance throughout most cases and they are consistent for different sample sizes. For Case Ⅰ, since the error distribution is normal, the performance of each estimate mainly depends on their efficiency. In this case, the OLS has the smallest MSE and MAD which is reasonable since under normal errors OLS is the best estimate; M_H, M_T, MM, and REWLSE have similar MSE to OLS, due to their high efficiency property; LMS, LTS, and S have relative larger MSE due to their low efficiency property. If the efficiency of some robust $\hat{\beta}^R$ relative to $\hat{\beta}^{OLS}$ is defined as the ratio of MSE of $\hat{\beta}^{OLS}$ to MSE of $\hat{\beta}^R$, the efficiencies of $\hat{\beta}_0^{LMS}$ and $\hat{\beta}_0^{LTS}$ relative to $\hat{\beta}_0^{OLS}$ do not exceed 28% for both $\hat{\beta}_0$ and $\hat{\beta}_1$. The efficiencies of $\hat{\beta}_0^S$ relative to $\hat{\beta}_0^{OLS}$ are between 32.38% and 41.60%, but the best efficiency of $\hat{\beta}_1^S$ relative to $\hat{\beta}_0^{OLS}$ is 29.5%. The efficiencies of other methods are much higher. MM, REWLSE, M_H, and M_T have smaller RB than those of LMS, LTS and S estimators. For Case Ⅱ, M_H, M_T, $\mathrm{M_M}$, and REWLSE work better than other estimates in terms of MSE and MAD. For Case Ⅲ, OLS has much larger MSE than other robust estimators; M_H, M_T, MM, REWLSE and S have similar MSE, RB and MAD. For Case Ⅳ, M_H, M_T, MM, and REWLSE have smaller MSE and MAD than others. From Case Ⅴ, we can see that when the data contain outliers in the y-direction, OLS is much worse than any other robust estimates; MM, REWLSE, and M_T are better than other robust estimators. The reason why M_T can perform better than M_H is that Tukey's bisquare function can completely remove the effect of large outliers while Huber's ψ function can only reduce the effect of large outliers. Finally for Case Ⅵ, since there are high leverage outliers, similar to OLS, both M_T and M_H perform poorly; MM and REWLSE work better than other robust estimates.

Table 1. 1 ***Comparison of Different Estimates for Example* 1. 1 *with* $n=20$**

	OLS	M_H	M_T	LMS	LTS	S	MM	REWLSE
				Case Ⅰ: $\varepsilon \sim N(0, 1)$				
MSE($\hat{\beta}_0$)	0. 0644	0. 0679	0. 0685	0. 2536	0. 2332	0. 1548	0. 0679	0. 0654
RB($\hat{\beta}_0$)	-0. 0121	-0. 0073	0. 0011	-0. 0093	-0. 0482	-0. 0045	0. 0032	0. 0105
MAD($\hat{\beta}_0$)	0. 1663	0. 1570	0. 1473	0. 3669	0. 3780	0. 2519	0. 1533	0. 1593
MSE($\hat{\beta}_1$)	0. 0544	0. 0578	0. 0584	0. 3117	0. 2455	0. 1843	0. 0572	0. 0563
RB($\hat{\beta}_1$)	-0. 0021	-0. 0060	-0. 0145	-0. 0002	0. 0513	0. 0247	0. 0001	0. 0041
MAD($\hat{\beta}_1$)	0. 1527	0. 1574	0. 1543	0. 3159	0. 3129	0. 2667	0. 1517	0. 1527
				Case Ⅱ: $\varepsilon \sim t_3$				
MSE($\hat{\beta}_0$)	0. 1337	0. 0804	0. 0829	0. 2374	0. 2259	0. 1293	0. 0825	0. 0867
RB($\hat{\beta}_0$)	-0. 0399	-0. 0151	0. 0064	-0. 0052	0. 0128	0. 0122	-0. 0055	0. 0082
MAD($\hat{\beta}_0$)	0. 2353	0. 1672	0. 1746	0. 3030	0. 2925	0. 2357	0. 1754	0. 1881
MSE($\hat{\beta}_1$)	0. 1867	0. 0947	0. 0952	0. 2924	0. 2836	0. 1739	0. 0902	0. 1007
RB($\hat{\beta}_1$)	0. 0251	-0. 0013	-0. 0160	0. 0223	0. 0416	0. 0055	-0. 0097	-0. 0037
MAD($\hat{\beta}_1$)	0. 1994	0. 1744	0. 1818	0. 3598	0. 3353	0. 2496	0. 1783	0. 1965
				Case Ⅲ: $\varepsilon \sim t_1$				
MSE($\hat{\beta}_0$)	2201. 5	0. 2245	0. 1655	0. 2347	0. 2388	0. 1752	0. 1764	0. 1727
RB($\hat{\beta}_0$)	0. 0979	0. 0227	0. 0224	0. 0253	0. 0095	0. 0053	0. 0064	-0. 0053
MAD($\hat{\beta}_0$)	0. 8252	0. 2991	0. 2553	0. 2801	0. 2876	0. 2471	0. 2646	0. 3095
MSE($\hat{\beta}_1$)	810. 20	0. 3723	0. 2192	0. 3962	0. 4527	0. 2209	0. 2313	0. 2272
RB($\hat{\beta}_1$)	-0. 0127	0. 0030	0. 0010	-0. 0924	-0. 1014	-0. 0541	-0. 0075	0. 0033
MAD($\hat{\beta}_1$)	0. 8094	0. 3142	0. 2659	0. 3550	0. 3046	0. 2586	0. 2808	0. 2853
				Case Ⅳ: $\varepsilon \sim 0.95N(0, 1) + 0.05N(0, 10^2)$				
MSE($\hat{\beta}_0$)	0. 3494	0. 0651	0. 0641	0. 2555	0. 2431	0. 1697	0. 0621	0. 0648
RB($\hat{\beta}_0$)	-0. 0899	-0. 0299	-0. 0171	-0. 0145	-0. 0459	-0. 0468	-0. 0162	-0. 0061
MAD($\hat{\beta}_0$)	0. 2961	0. 1652	0. 1763	0. 3273	0. 3424	0. 2846	0. 1696	0. 1676
MSE($\hat{\beta}_1$)	0. 3261	0. 0677	0. 0601	0. 3433	0. 3164	0. 1558	0. 0585	0. 0527

Table 1.1 *Comparison of Different Estimates for Example* 1.1 *with* $n=20$ (Continued table)

	OLS	M_H	M_T	LMS	LTS	S	MM	REWLSE
RB($\hat{\beta}_1$)	-0.0364	-0.0200	-0.0178	0.0161	-0.0069	-0.0055	-0.0169	-0.0214
MAD($\hat{\beta}_1$)	0.2344	0.1536	0.1477	0.2996	0.2996	0.2492	0.1476	0.1556
			Case Ⅴ: $\varepsilon \sim N$ (0, 1) with outliers in y direction					
MSE($\hat{\beta}_0$)	9.4097	0.0966	0.0483	0.2238	0.1943	0.1306	0.0473	0.0423
RB($\hat{\beta}_0$)	3.0130	0.2288	0.0267	-0.0356	-0.0230	0.0010	0.0267	0.0250
MAD($\hat{\beta}_0$)	3.0130	0.2385	0.1330	0.3024	0.3062	0.2362	0.1290	0.1425
MSE($\hat{\beta}_1$)	4.9846	0.0912	0.0608	0.3241	0.2599	0.1847	0.0603	0.0603
RB($\hat{\beta}_1$)	0.0443	0.0126	-0.0045	0.0036	-0.0433	-0.0020	0.0013	-0.0004
MAD($\hat{\beta}_1$)	1.3387	0.1793	0.1458	0.3305	0.3065	0.2313	0.1502	0.1500
			Case Ⅵ: $\varepsilon \sim N$ (0, 1) with high leverage outliers					
MSE($\hat{\beta}_0$)	0.7718	0.8123	0.8293	0.2228	0.2021	0.1370	0.0718	0.0717
RB($\hat{\beta}_0$)	0.2579	0.2347	0.2790	0.0491	0.0303	0.0084	0.0030	0.0009
MAD($\hat{\beta}_0$)	0.5265	0.5485	0.5590	0.2981	0.2892	0.2422	0.1540	0.1559
MSE($\hat{\beta}_1$)	13.397	13.738	13.878	0.3435	0.2287	0.1661	0.0772	0.0731
RB($\hat{\beta}_1$)	3.6644	3.7123	3.7294	0.0129	0.0500	-0.0280	0.0126	0.0129
MAD($\hat{\beta}_1$)	3.6644	3.7123	3.7294	0.2944	0.2882	0.2708	0.1603	0.1588

Table 1.2 *Comparison of Different Estimates for Example* 1.1 *with* $n=100$

TRUE	OLS	M_H	M_T	LMS	LTS	S	MM	REWLSE
				Case Ⅰ: $\varepsilon \sim N$ (0, 1)				
MSE($\hat{\beta}_0$)	0.0102	0.0117	0.0118	0.0675	0.0728	0.0315	0.0118	0.0109
RB($\hat{\beta}_0$)	0.0049	-0.0037	-0.0031	0.0084	0.0357	0.0272	-0.0015	-0.0039
MAD($\hat{\beta}_0$)	0.0724	0.0737	0.0730	0.1819	0.1829	0.1225	0.0734	0.0727
MSE($\hat{\beta}_1$)	0.0105	0.0112	0.0114	0.0610	0.0762	0.0367	0.0114	0.0115
RB($\hat{\beta}_1$)	-0.0108	0.0002	-0.0032	-0.0167	-0.0028	-0.0007	-0.0047	-0.0035
MAD($\hat{\beta}_1$)	0.0717	0.0700	0.0759	0.1670	0.1785	0.1188	0.0762	0.0761

Table 1.2 ***Comparison of Different Estimates for Example* 1.1 *with n* =100 (Continued table)**

TRUE	OLS	M_H	M_T	LMS	LTS	S	MM	REWLSE
				Case Ⅱ: $\varepsilon \sim t_3$				
MSE($\hat{\beta}_0$)	0.0355	0.0168	0.0164	0.0600	0.0612	0.0284	0.0165	0.0166
RB($\hat{\beta}_0$)	−0.0255	−0.0018	0.0037	0.0220	0.0051	0.0167	0.0052	0.0037
MAD($\hat{\beta}_0$)	0.1133	0.0868	0.0736	0.1703	0.1849	0.1010	0.0767	0.0819
MSE($\hat{\beta}_1$)	0.0394	0.0196	0.0189	0.0589	0.0661	0.0324	0.0187	0.0190
RB($\hat{\beta}_1$)	−0.0363	−0.0293	−0.0335	−0.0229	−0.0087	−0.0265	−0.0309	−0.0278
MAD($\hat{\beta}_1$)	0.1235	0.0896	0.0927	0.1501	0.1625	0.1167	0.0892	0.1027
				Case Ⅲ: $\varepsilon \sim t_1$				
MSE($\hat{\beta}_0$)	8810.4	0.0405	0.0327	0.0713	0.0486	0.0304	0.0342	0.0339
RB($\hat{\beta}_0$)	0.0687	−0.0213	0.0023	−0.0230	−0.0107	−0.0165	0.0020	−0.0250
MAD($\hat{\beta}_0$)	1.1700	0.1385	0.1216	0.1720	0.1374	0.1040	0.1245	0.1235
MSE($\hat{\beta}_1$)	27954	0.0396	0.0298	0.0667	0.0466	0.0337	0.0300	0.0302
RB($\hat{\beta}_1$)	−0.0347	0.0117	0.0011	0.0225	−0.0044	0.0188	0.0003	0.0006
MAD($\hat{\beta}_1$)	1.1136	0.1300	0.1078	0.1556	0.1317	0.1083	0.1078	0.1060
				Case Ⅳ: $\varepsilon \sim 0.95N(0, 1) + 0.05N(0, 10^2)$				
MSE($\hat{\beta}_0$)	49.49	0.0139	0.0131	0.0757	0.0752	0.0344	0.0130	0.0136
RB($\hat{\beta}_0$)	−0.0365	0.0036	−0.0012	0.0607	0.0293	0.0257	−0.0011	0.0002
MAD($\hat{\beta}_0$)	0.1742	0.0723	0.0702	0.1990	0.1848	0.1289	0.0691	0.0716
MSE($\hat{\beta}_1$)	1.4060	0.0125	0.0108	0.0576	0.0698	0.0291	0.0112	0.0112
RB($\hat{\beta}_1$)	−0.0209	−0.0072	−0.0073	0.0150	−0.0009	0.0012	−0.0092	−0.0140
MAD($\hat{\beta}_1$)	0.1448	0.0725	0.0690	0.1675	0.1552	0.1070	0.0704	0.0734
				Case Ⅴ: $\varepsilon \sim N(0, 1)$ with outliers in y direction				
MSE($\hat{\beta}_0$)	8.9967	0.0500	0.0123	0.0701	0.0724	0.0313	0.0122	0.0125
RB($\hat{\beta}_0$)	2.9916	0.1842	−0.0101	−0.0023	−0.0174	−0.0230	−0.0114	−0.0127
MAD($\hat{\beta}_0$)	2.9916	0.1842	0.0838	0.1827	0.1889	0.1228	0.0782	0.0843
MSE($\hat{\beta}_1$)	0.9034	0.0169	0.0115	0.0641	0.0708	0.0305	0.0113	0.0117

Table 1.2 ***Comparison of Different Estimates for Example* 1.1 *with n* =100 (Continued table)**

TRUE	OLS	M_H	M_T	LMS	LTS	S	MM	REWLSE
RB($\hat{\beta}_1$)	-0.0817	0.0101	0.0090	0.0105	0.0003	0.0149	0.0068	0.0001
MAD($\hat{\beta}_1$)	0.6839	0.0819	0.0734	0.1725	0.1926	0.1060	0.0734	0.0759
Case Ⅵ: $\varepsilon\sim N(0, 1)$ with high leverage outliers								
MSE($\hat{\beta}_0$)	0.2942	0.3116	0.3198	0.0678	0.0615	0.0296	0.0123	0.0126
RB($\hat{\beta}_0$)	0.3942	0.3645	0.3709	-0.0020	0.0094	-0.0134	-0.0028	-0.0048
MAD($\hat{\beta}_0$)	0.4118	0.4124	0.4210	0.1694	0.1664	0.1165	0.0768	0.0752
MSE($\hat{\beta}_1$)	13.269	13.621	13.740	0.0643	0.0616	0.0346	0.0119	0.0117
RB($\hat{\beta}_1$)	3.6445	3.6860	3.7033	0.0010	-0.0321	0.0054	0.0101	0.0115
MAD($\hat{\beta}_1$)	3.6445	3.6860	3.7033	0.1786	0.1647	0.1117	0.0634	0.0660

In order to better compare the performance of different methods, Figure 1.1 shows the plot of their MSE versus each case for the intercept (left side) and slope

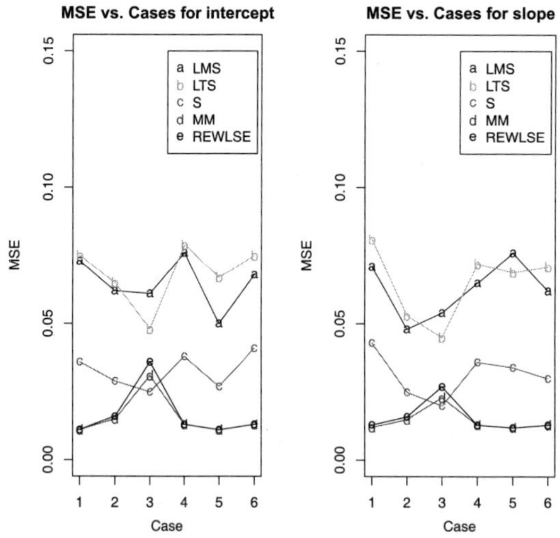

Figure 1.1 ***Plot of MSE of intercept (left) and slope (right) estimates vs. different cases for LMS, LTS, S, MM, and REWLSE, for Example* 1.1 *when n* =100**

(right side) parameters for Example 1.1 when sample size $n=100$. Since the lines for LTS and LMS are above the other lines, S, MM, and REWLSE of the intercept and slopes outperform LTS and LMS estimates throughout all six cases. In addition, the S estimate has similar performance to MM and REWLSE when the error density of ε is Cauchy distribution. However, MM and REWLSE perform better than S-estimates in other five cases. Furthermore, the lines for MM and REWLSE almost overlap for all six cases. It shows that MM and REWLSE are the overall best approaches in robust regression.

Example 1.2: Samples $\{(x_1, y_1), \ldots, (x_n, y_n)\}$ are generated from the model

$$Y=X_1+X_2+X_3+\varepsilon$$

where $X_i \sim N(0, 1)$, $i=1, 2, 3$ and X_is are independent. We consider the following six cases for the error density of ε:

Case Ⅰ: $\varepsilon \sim N(0, 1)$-standard normal distribution.

Case Ⅱ: $\varepsilon \sim t_3$-t-distribution with degrees of freedom 3.

Case Ⅲ: $\varepsilon \sim t_1$-t-distribution with degrees of freedom 1 (Cauchy distribution).

Case Ⅳ: $\varepsilon \sim 0.95N(0, 1) + 0.05N(0, 10^2)$-contaminated normal mixture.

Case Ⅴ: $\varepsilon \sim N(0, 1)$ with 10% identical outliers in y direction (where we let the first 10% of y's equal to 30).

Case Ⅵ: $\varepsilon \sim N(0, 1)$ with 10% identical high leverage outliers being $X_1=10$, $X_2=10$, $X_3=10$, and $Y=50$.

Table 1.3 and Table 1.4 show MSE, RB and MAD of the parameter estimates of each estimation method for sample size $n = 20$ and $n = 100$, respectively. Figure 1. 2 shows the plot of their MSE versus each case for three slopes and the intercept parameters with sample size $n = 100$. The results in Example 1. 2 tell similar stories to Example 1. 1. In summary, MM and REWLSE have the overall best performance; OLS only works well when there are no outliers since it is very sensitive to outliers; M-estimates (M_H and M_T) work well if the outliers are in y direction but are also sensitive to the high leverage outliers.

Table 1. 3 ***Comparison of Different Estimates for Example* 1. 2 *with* $n = 20$**

TRUE	OLS	M_H	M_T	LMS	LTS	S	MM	REWLSE
				Case Ⅰ: $\varepsilon \sim N(0, 1)$				
MSE($\hat{\beta}_0$)	0. 0574	0. 0612	0. 0660	0. 3700	0. 2544	0. 1870	0. 0622	0. 0613
RB($\hat{\beta}_0$)	0. 0136	0. 0081	0. 0053	−0. 0068	−0. 0673	−0. 0435	−0. 0033	−0. 0016
MAD($\hat{\beta}_0$)	0. 1707	0. 1879	0. 1955	0. 4187	0. 3356	0. 2912	0. 1889	0. 1871
MSE($\hat{\beta}_1$)	0. 0670	0. 0732	0. 0866	0. 4648	0. 3008	0. 2550	0. 0776	0. 0730
RB($\hat{\beta}_1$)	−0. 0259	−0. 0253	−0. 0256	−0. 0359	0. 0227	−0. 0100	−0. 0228	−0. 0252
MAD($\hat{\beta}_1$)	0. 1723	0. 1824	0. 1955	0. 4119	0. 3317	0. 2857	0. 1795	0. 1728
MSE($\hat{\beta}_2$)	0. 0642	0. 0624	0. 0667	0. 4648	0. 3275	0. 2544	0. 0647	0. 0648
RB($\hat{\beta}_2$)	0. 0086	0. 0026	−0. 0010	0. 0173	0. 0365	−0. 0386	0. 0070	0. 0014
MAD($\hat{\beta}_2$)	0. 1760	0. 1664	0. 1643	0. 4029	0. 3535	0. 2994	0. 1704	0. 1748
MSE($\hat{\beta}_3$)	0. 0706	0. 0751	0. 0829	0. 4202	0. 3052	0. 2283	0. 0789	0. 0748
RB($\hat{\beta}_3$)	−0. 0099	0. 0084	0. 0026	−0. 1113	−0. 0516	−0. 0244	0. 0001	−0. 0057
MAD($\hat{\beta}_3$)	0. 1680	0. 1761	0. 1859	0. 3983	0. 3461	0. 2765	0. 1802	0. 1718
				Case Ⅱ: $\varepsilon \sim t_3$				
MSE($\hat{\beta}_0$)	0. 1736	0. 0927	0. 0919	0. 3861	0. 2972	0. 1970	0. 0904	0. 0949
RB($\hat{\beta}_0$)	0. 0416	0. 0414	0. 0292	0. 1139	0. 1070	0. 0601	0. 0300	0. 0157

Table 1.3 ***Comparison of Different Estimates for Example* 1.2 *with* n=20 (Continued table)**

TRUE	OLS	M_H	M_T	LMS	LTS	S	MM	REWLSE
MAD($\hat{\beta}_0$)	0.2544	0.2274	0.2110	0.4091	0.3620	0.2878	0.2024	0.1970
MSE($\hat{\beta}_1$)	0.1959	0.1545	0.1560	0.5253	0.4219	0.2548	0.1557	0.1535
RB($\hat{\beta}_1$)	0.0248	0.0135	0.0214	-0.0811	-0.0361	0.0100	0.0239	0.0150
MAD($\hat{\beta}_1$)	0.2470	0.2329	0.2376	0.4159	0.3234	0.3009	0.2300	0.2317
MSE($\hat{\beta}_2$)	0.2878	0.1690	0.1638	0.6129	0.3644	0.2437	0.1580	0.1614
RB($\hat{\beta}_2$)	-0.0509	-0.0188	-0.0179	-0.0465	-0.0039	0.0166	-0.0130	0.0082
MAD($\hat{\beta}_2$)	0.2743	0.2432	0.2428	0.4551	0.3399	0.2901	0.2351	0.2399
MSE($\hat{\beta}_3$)	0.2566	0.1383	0.1359	0.5529	0.4124	0.3023	0.1404	0.1380
RB($\hat{\beta}_3$)	0.0433	0.0427	0.0071	0.0568	0.0804	-0.0006	0.0010	-0.0013
MAD($\hat{\beta}_3$)	0.2517	0.2140	0.2364	0.4065	0.3745	0.2883	0.2346	0.2251
Case Ⅲ: $\varepsilon \sim t_1$								
MSE($\hat{\beta}_0$)	33406	0.4019	0.3217	0.8503	0.4983	0.3847	0.3298	0.3258
RB($\hat{\beta}_0$)	-0.2515	-0.1037	-0.0377	-0.0176	-0.0483	-0.0501	-0.0545	-0.0510
MAD($\hat{\beta}_0$)	0.9030	0.3429	0.3327	0.4471	0.3553	0.3459	0.2961	0.3172
MSE($\hat{\beta}_1$)	2069.0	0.4238	0.3401	1.1884	0.6947	0.3958	0.3393	0.3424
RB($\hat{\beta}_1$)	-0.1555	-0.0913	-0.0909	0.0056	-0.0019	0.0112	-0.0561	-0.0483
MAD($\hat{\beta}_1$)	1.0759	0.3655	0.3286	0.4701	0.3801	0.3219	0.3148	0.3274
MSE($\hat{\beta}_2$)	3188.0	0.6564	0.4883	0.9200	0.6419	0.4245	0.4616	0.4526
RB($\hat{\beta}_2$)	0.1034	0.0541	-0.0091	0.0396	0.0210	-0.0583	-0.0606	0.0026
MAD($\hat{\beta}_2$)	0.9409	0.4642	0.3538	0.4468	0.4039	0.3362	0.3965	0.3491
MSE($\hat{\beta}_3$)	1774.0	0.5386	0.4867	1.0639	0.6637	0.4577	0.4923	0.4951
RB($\hat{\beta}_3$)	0.0863	0.0229	-0.0303	-0.1065	-0.0568	-0.0215	-0.0155	-0.0107
MAD($\hat{\beta}_3$)	0.8467	0.3781	0.3606	0.4403	0.3913	0.3599	0.3615	0.3667
Case Ⅳ: $\varepsilon \sim 0.95N(0, 1) + 0.05N(0, 10^2)$								
MSE($\hat{\beta}_0$)	0.4870	0.0919	0.0754	0.3972	0.2812	0.1929	0.0755	0.0721
RB($\hat{\beta}_0$)	0.0447	0.0215	0.0108	-0.0390	0.0195	-0.0209	0.0184	0.0216

Table 1. 3 ***Comparison of Different Estimates for Example* 1. 2 *with* $n=20$ (Continued table)**

TRUE	OLS	M_H	M_T	LMS	LTS	S	MM	REWLSE
MAD($\hat{\beta}_0$)	0. 2980	0. 2172	0. 1838	0. 4019	0. 3822	0. 3052	0. 1875	0. 1854
MSE($\hat{\beta}_1$)	0. 4734	0. 1240	0. 0983	0. 5312	0. 3103	0. 2023	0. 0976	0. 1009
RB($\hat{\beta}_1$)	-0. 0162	-0. 0055	0. 0090	-0. 0115	-0. 0055	0. 0060	0. 0003	0. 0046
MAD($\hat{\beta}_1$)	0. 3040	0. 2209	0. 1990	0. 3813	0. 3508	0. 2743	0. 2090	0. 2072
MSE($\hat{\beta}_2$)	0. 6811	0. 1069	0. 0861	0. 4851	0. 3101	0. 2033	0. 0939	0. 1072
RB($\hat{\beta}_2$)	-0. 0642	-0. 0265	-0. 0146	-0. 0774	-0. 0313	-0. 0361	-0. 0177	-0. 0092
MAD($\hat{\beta}_2$)	0. 2557	0. 1722	0. 1594	0. 4231	0. 3303	0. 2592	0. 1612	0. 1805
MSE($\hat{\beta}_3$)	0. 5937	0. 1055	0. 0798	0. 5802	0. 3336	0. 2359	0. 0773	0. 0736
RB($\hat{\beta}_3$)	-0. 0083	-0. 0033	-0. 0126	0. 0150	-0. 0341	-0. 0323	-0. 0065	-0. 0055
MAD($\hat{\beta}_3$)	0. 3409	0. 2131	0. 1928	0. 3499	0. 3382	0. 2876	0. 1891	0. 2133
Case Ⅴ: $\varepsilon \sim N(0, 1)$ with outliers in y direction								
MSE($\hat{\beta}_0$)	9. 8831	0. 1470	0. 0784	0. 3489	0. 2487	0. 1643	0. 0735	0. 0739
RB($\hat{\beta}_0$)	2. 9410	0. 2160	0. 0030	-0. 0628	-0. 0392	-0. 0053	0. 0025	0. 0110
MAD($\hat{\beta}_0$)	2. 9410	0. 2655	0. 1807	0. 4015	0. 3288	0. 2530	0. 1818	0. 1760
MSE($\hat{\beta}_1$)	4. 6973	0. 0918	0. 0721	0. 4046	0. 2875	0. 2069	0. 0668	0. 0697
RB($\hat{\beta}_1$)	-0. 1366	-0. 0105	-0. 0385	0. 0102	0. 0221	-0. 0005	-0. 0243	-0. 0134
MAD($\hat{\beta}_1$)	1. 3713	0. 1846	0. 1844	0. 3920	0. 3655	0. 3075	0. 1775	0. 1846
MSE($\hat{\beta}_2$)	6. 1743	0. 1736	0. 1001	0. 5268	0. 2976	0. 2143	0. 0936	0. 0955
RB($\hat{\beta}_2$)	-0. 2108	-0. 0370	-0. 0079	0. 0435	0. 0053	-0. 0081	-0. 0117	-0. 0304
MAD($\hat{\beta}_2$)	1. 667	0. 2181	0. 1775	0. 3829	0. 3649	0. 2845	0. 1833	0. 1800
MSE($\hat{\beta}_3$)	5. 5139	0. 1331	0. 0781	0. 3906	0. 2896	0. 2088	0. 0746	0. 0855
RB($\hat{\beta}_3$)	-0. 2013	-0. 0291	-0. 0322	-0. 0773	-0. 0335	-0. 0129	-0. 0322	-0. 0337
MAD($\hat{\beta}_3$)	1. 5581	0. 2392	0. 1874	0. 3885	0. 3493	0. 2791	0. 1889	0. 1868
Case Ⅵ: $\varepsilon \sim N(0, 1)$ with high leverage outliers								
MSE($\hat{\beta}_0$)	0. 9099	0. 9893	1. 0550	0. 3347	0. 29325	0. 1537	0. 0685	0. 0679
RB($\hat{\beta}_0$)	0. 1594	0. 1545	0. 1863	0. 0552	0. 0736	0. 0472	-0. 0014	0. 0098

Table 1.3 ***Comparison of Different Estimates for Example* 1.2 *with* $n=20$ (Continued table)**

TRUE	OLS	M_H	M_T	LMS	LTS	S	MM	REWLSE
MAD($\hat{\beta}_0$)	0.6569	0.7016	0.7226	0.3627	0.3188	0.2750	0.1642	0.1783
MSE($\hat{\beta}_1$)	13.504	13.770	13.8048	0.4387	0.3476	0.2132	0.0980	0.1090
RB($\hat{\beta}_1$)	3.6694	3.7134	3.7218	-0.0265	0.0247	-0.0354	-0.0493	-0.0466
MAD($\hat{\beta}_1$)	3.6694	3.7134	3.7218	0.3999	0.3257	0.2841	0.1791	0.1938
MSE($\hat{\beta}_2$)	0.8393	0.9009	0.9702	0.3716	0.2452	0.1608	0.0797	0.0767
RB($\hat{\beta}_2$)	-0.2032	-0.1788	-0.1737	-0.0156	-0.0173	0.0307	-0.0401	-0.0091
MAD($\hat{\beta}_2$)	0.6166	0.6425	0.6462	0.3405	0.2932	0.2346	0.1693	0.1795
MSE($\hat{\beta}_3$)	0.7862	0.8487	0.9068	0.3964	0.3133	0.1900	0.0839	0.0919
RB($\hat{\beta}_3$)	-0.1069	-0.1278	-0.1327	-0.1022	-0.0969	-0.0599	-0.0260	-0.0374
MAD($\hat{\beta}_3$)	0.6878	0.6706	0.7013	0.3196	0.3171	0.2757	0.1871	0.1830

Table 1.4 ***Comparison of Different Estimates for Example* 1.2 *with* $n=100$**

TRUE	OLS	M_H	M_T	LMS	LTS	S	MM	REWLSE
				Case Ⅰ: $\varepsilon \sim N(0, 1)$				
MSE($\hat{\beta}_0$)	0.0086	0.0090	0.0090	0.0750	0.0778	0.0398	0.0089	0.0089
RB($\hat{\beta}_0$)	0.0119	0.0089	0.0087	0.0127	0.0245	0.0100	0.0076	0.0080
MAD($\hat{\beta}_0$)	0.0669	0.0684	0.0677	0.1862	0.1995	0.1329	0.0682	0.0649
MSE($\hat{\beta}_1$)	0.0102	0.0111	0.0111	0.0660	0.0753	0.0459	0.0111	0.0111
RB($\hat{\beta}_1$)	-0.0025	-0.0062	-0.0053	-0.0016	-0.0077	-0.0222	-0.0067	-0.0065
MAD($\hat{\beta}_1$)	0.0688	0.0704	0.0751	0.1559	0.1875	0.1653	0.0761	0.0761
MSE($\hat{\beta}_2$)	0.0118	0.0121	0.0124	0.0682	0.0786	0.0488	0.0123	0.0120
RB($\hat{\beta}_2$)	0.0084	0.0072	0.0106	0.0395	0.0055	-0.0150	0.0121	0.0045
MAD($\hat{\beta}_2$)	0.0727	0.0710	0.0694	0.1742	0.1718	0.1570	0.0710	0.0713
MSE($\hat{\beta}_3$)	0.0098	0.0102	0.0101	0.0621	0.0564	0.0372	0.0101	0.0101
RB($\hat{\beta}_3$)	0.0073	0.0041	0.0065	-0.0005	-0.0019	-0.0026	0.0054	0.0048
MAD($\hat{\beta}_3$)	0.0705	0.0702	0.0699	0.1714	0.1508	0.1172	0.0708	0.0703

Table 1.4 ***Comparison of Different Estimates for Example 1.2 with n = 100*** **(Continued table)**

TRUE	OLS	M_H	M_T	LMS	LTS	S	MM	REWLSE
				Case Ⅱ: $\varepsilon \sim t_3$				
MSE($\hat{\beta}_0$)	0.0214	0.0134	0.0137	0.0546	0.0554	0.0326	0.0138	0.0153
RB($\hat{\beta}_0$)	0.0066	0.0019	−0.0013	−0.0362	−0.0202	−0.0162	−0.0002	0.0010
MAD($\hat{\beta}_0$)	0.1103	0.0815	0.0787	0.1319	0.1640	0.1324	0.0790	0.0811
MSE($\hat{\beta}_1$)	0.0271	0.0139	0.0139	0.0767	0.0703	0.0455	0.0137	0.0140
RB($\hat{\beta}_1$)	−0.0257	−0.0116	−0.0149	−0.034	0.0110	−0.0023	−0.0144	−0.0097
MAD($\hat{\beta}_1$)	0.1135	0.0852	0.0818	0.1658	0.1618	0.1272	0.0781	0.0852
MSE($\hat{\beta}_2$)	0.0320	0.0185	0.0177	0.0644	0.0601	0.0416	0.0181	0.0184
RB($\hat{\beta}_2$)	0.0044	0.0170	0.0311	0.0078	0.0437	0.0196	0.0293	0.0250
MAD($\hat{\beta}_2$)	0.1245	0.0901	0.0858	0.1618	0.1688	0.1369	0.0865	0.0846
MSE($\hat{\beta}_3$)	0.0311	0.0190	0.0190	0.0709	0.0658	0.0451	0.0188	0.0203
RB($\hat{\beta}_3$)	0.0022	−0.0037	0.0055	−0.0014	−0.0057	0.0235	0.0025	0.0038
MAD($\hat{\beta}_3$)	0.1231	0.0968	0.0955	0.1731	0.1600	0.1442	0.0952	0.1002
				Case Ⅲ: $\varepsilon \sim t_1$				
MSE($\hat{\beta}_0$)	178.66	0.0378	0.0301	0.0624	0.0455	0.0305	0.0311	0.0330
RB($\hat{\beta}_0$)	−0.0912	−0.0267	−0.0017	−0.0239	−0.0108	−0.0169	−0.0087	−0.0124
MAD($\hat{\beta}_0$)	0.9190	0.1154	0.1205	0.1607	0.1351	0.1170	0.1139	0.1185
MSE($\hat{\beta}_1$)	72.519	0.0445	0.0333	0.0693	0.0634	0.0528	0.0359	0.0359
RB($\hat{\beta}_1$)	−0.1049	−0.0095	−0.0007	−0.0002	−0.0089	−0.0068	0.0021	−0.0002
MAD($\hat{\beta}_1$)	0.7958	0.1288	0.1218	0.1686	0.1418	0.1494	0.1292	0.1349
MSE($\hat{\beta}_2$)	198.19	0.0370	0.0342	0.0708	0.0658	0.0535	0.0352	0.0377
RB($\hat{\beta}_2$)	0.0636	0.0200	0.0035	−0.0147	0.0183	−0.0044	0.0049	0.0068
MAD($\hat{\beta}_2$)	0.6610	0.1184	0.1118	0.1645	0.1449	0.1573	0.1143	0.1268
MSE($\hat{\beta}_3$)	68.1196	0.0389	0.0323	0.0579	0.0481	0.0446	0.0338	0.0325
RB($\hat{\beta}_3$)	−0.1051	−0.0111	−0.0265	−0.0210	−0.0284	−0.0221	−0.0283	−0.0293
MAD($\hat{\beta}_3$)	0.812	0.1204	0.1232	0.1598	0.1339	0.1389	0.1294	0.1194

Table 1.4 ***Comparison of Different Estimates for Example*** **1. 2** ***with n*** **=100 (Continued table)**

TRUE	OLS	M_H	M_T	LMS	LTS	S	MM	REWLSE
			Case Ⅳ: $\varepsilon \sim 0.95N(0, 1) + 0.05N(0, 10^2)$					
MSE($\hat{\beta}_0$)	0.0600	0.0136	0.0120	0.0666	0.0697	0.0365	0.0119	0.0119
RB($\hat{\beta}_0$)	-0.0074	0.0002	-0.0004	-0.0182	-0.0302	-0.0182	-0.0026	-0.0037
MAD($\hat{\beta}_0$)	0.1605	0.0811	0.0756	0.1874	0.1751	0.1370	0.0768	0.0771
MSE($\hat{\beta}_1$)	0.0638	0.0162	0.01499	0.0661	0.0786	0.0499	0.0150	0.0154
RB($\hat{\beta}_1$)	-0.0306	0.0139	0.0106	0.0049	0.0037	0.0100	0.0118	0.0190
MAD($\hat{\beta}_1$)	0.1640	0.0819	0.0832	0.1667	0.1835	0.1488	0.0829	0.0857
MSE($\hat{\beta}_2$)	0.0522	0.0128	0.0120	0.0621	0.0679	0.0368	0.0120	0.0130
RB($\hat{\beta}_2$)	0.0278	0.0125	0.0012	-0.0112	0.0070	-0.0040	0.0017	0.0040
MAD($\hat{\beta}_2$)	0.1359	0.07363	0.0709	0.1876	0.1647	0.1328	0.0706	0.0749
MSE($\hat{\beta}_3$)	0.0782	0.0170	0.0154	0.0706	0.0746	0.0422	0.0153	0.0156
RB($\hat{\beta}_3$)	-0.0063	-0.0031	0.0038	0.0036	0.0241	-0.0191	0.0048	0.0074
MAD($\hat{\beta}_3$)	0.1804	0.0759	0.0823	0.1651	0.1791	0.1373	0.0824	0.0796
			Case Ⅴ: $\varepsilon \sim N(0, 1)$ with outliers in y direction					
MSE($\hat{\beta}_0$)	9.0524	0.0571	0.0116	0.0669	0.0776	0.0457	0.0115	0.0115
RB($\hat{\beta}_0$)	3.0111	0.2139	0.0153	0.0299	0.0341	0.0315	0.0124	0.0088
MAD($\hat{\beta}_0$)	3.0111	0.2139	0.0698	0.1735	0.1980	0.1384	0.0693	0.0708
MSE($\hat{\beta}_1$)	0.9323	0.0137	0.0110	0.0574	0.0595	0.0378	0.0108	0.0116
RB($\hat{\beta}_1$)	-0.1124	0.0084	0.0098	0.0078	0.0304	0.0118	0.0104	0.0089
MAD($\hat{\beta}_1$)	0.6464	0.0808	0.0712	0.1487	0.1520	0.1182	0.0685	0.0749
MSE($\hat{\beta}_2$)	0.8554	0.01487	0.0121	0.0596	0.0675	0.0387	0.0120	0.0123
RB($\hat{\beta}_2$)	-0.1011	-0.0224	-0.0161	-0.0108	0.0105	-0.0219	-0.0133	-0.0162
MAD($\hat{\beta}_2$)	0.6016	0.0822	0.0724	0.1605	0.1849	0.1479	0.0715	0.0695
MSE($\hat{\beta}_3$)	0.8373	0.0161	0.0125	0.0706	0.0618	0.0378	0.0124	0.0130
RB($\hat{\beta}_3$)	-0.0809	0.0031	-0.0035	-0.0020	-0.0401	-0.0086	-0.0045	-0.0021
MAD($\hat{\beta}_3$)	0.5779	0.0823	0.0774	0.1688	0.1731	0.1381	0.0774	0.0756

Table 1.4 ***Comparison of Different Estimates for Example 1.2 with n = 100 (Continued table)***

TRUE	OLS	M_H	M_T	LMS	LTS	S	MM	REWLSE
			Case Ⅵ: $\varepsilon \sim N(0, 1)$ with high leverage outliers					
MSE($\hat{\beta}_0$)	0.2092	0.2077	0.2095	0.0594	0.0707	0.0435	0.0111	0.0114
RB($\hat{\beta}_0$)	0.3088	0.2964	0.2991	−0.0003	−0.0203	−0.0215	−0.0107	−0.0131
MAD($\hat{\beta}_0$)	0.3358	0.3399	0.3386	0.1531	0.1871	0.1335	0.0694	0.0720
MSE($\hat{\beta}_1$)	13.303	13.683	13.726	0.0674	0.0759	0.0491	0.0114	0.0118
RB($\hat{\beta}_1$)	3.6473	3.6979	3.7026	0.0183	−0.0120	0.0132	−0.0051	−0.0047
MAD($\hat{\beta}_1$)	3.6473	3.6979	3.7026	0.1645	0.1842	0.1546	0.0758	0.0713
MSE($\hat{\beta}_2$)	0.1728	0.1812	0.1858	0.0640	0.0683	0.0373	0.0120	0.0124
RB($\hat{\beta}_2$)	−0.1141	−0.1184	−0.1281	0.0054	−0.0184	0.0069	−0.0147	−0.0132
MAD($\hat{\beta}_2$)	0.2689	0.2771	0.2796	0.1491	0.1676	0.1106	0.0712	0.0747
MSE($\hat{\beta}_3$)	0.1557	0.1592	0.1600	0.0641	0.0603	0.0342	0.0132	0.0137
RB($\hat{\beta}_3$)	−0.1083	−0.1104	−0.1088	−0.0239	−0.0222	0.0047	−0.0045	0.0034
MAD($\hat{\beta}_3$)	0.2561	0.2614	0.2637	0.1664	0.1564	0.1299	0.0724	0.0783

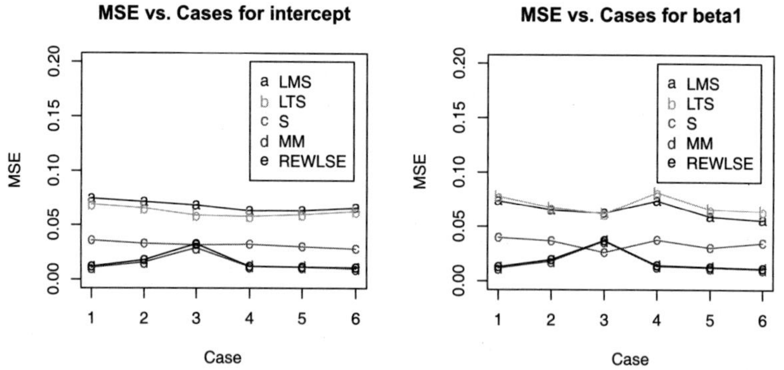

Figure 1.2 ***Plot of MSE of different regression parameter estimates vs. different cases for LMS, LTS, S, MM, and REWLSE, for Example 1.2 when n = 100***

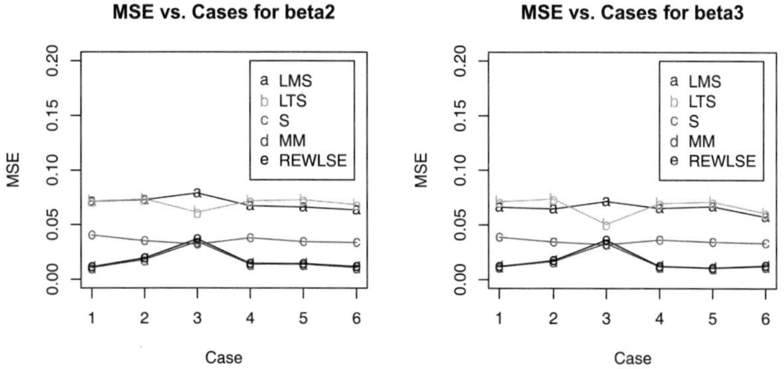

Figure 1.2 ***Plot of MSE of different regression parameter estimates vs. different cases for LMS, LTS, S, MM, and REWLSE, for Example* 1.2 *when n* =100 (Continued figure)**

Example 1.3: In order to compare the performance of outlier detection, we consider two cases: 5% and 10% high leverage outliers in the model. $n=100$ samples $\{(x_1, y_1), \ldots, (x_n, y_n)\}$ are generated from the model

$$Y=X_1+X_2+X_3+\gamma+\varepsilon$$

where γ is a vector, $\varepsilon \sim N(0, 1)$, $X_i \sim N(0, 1)$, $i=1, 2, 3$ and X_is are independent. We modify the first O rows of predictor matrix X to be $X_1=10$, $X_2=10$, $X_3=10$, where $O \in \{5, 10\}$. The first O rows of $\boldsymbol{\gamma}$ are randomly generated from a uniform distribution between 11 and 13, and the remaining $n-O$ rows of $\boldsymbol{\gamma}$ are all zeros. Therefore, the first O observations are high leverage outliers. In order to compare the performance of outlier detection of different methods, we use three benchmark proportions: M, S and JD (She and Owen, 2011). M is the mean masking probability (fraction of undetected true outliers), S denotes the mean swamping probability (fraction of good points labeled as outliers), and JD means the joint outlier detection rate (fraction of simulations with 0 masking). Ideally, $M\approx 0$, $S\approx 0$ and $JD\approx 100\%$.

As can be seen from Table 1.5, S and MM have relatively small probabilities of both masking and swamping, and LTS has smaller probability of masking but higher probability of swamping in the presence of 5% outliers. When the proportion of outliers increases to 10%, LMS, LTS, S, MM and REWLSE still have high joint identification rates which are larger than 60%. LMS has lowest masking probability but also has highest swamping probability. As expected, M_H and M_T have very high masking probabilities due to their sensitivity to high leverage outliers.

Table 1.5 ***Outlier detection results for Example*** **1.3**

	5% outliers			10% outliers		
	M	S	JD	M	S	JD
M_H	0.976	0.008	0.000	0.983	0.009	0.000
M_T	0.976	0.008	0.000	0.983	0.009	0.000
LMS	0.122	0.030	0.865	0.285	0.029	0.690
LTS	0.091	0.0249	0.900	0.332	0.025	0.645
S	0.088	0.008	0.905	0.365	0.008	0.625
MM	0.088	0.006	0.910	0.379	0.006	0.615
REWLSE	0.118	0.006	0.875	0.348	0.005	0.645

Example 1.4: We next use the modified data on wood specific gravity (Rousseeuw, 1984; Rousseeuw and Leroy, 1987; Olive and Hawkins, 2011) to compare OLS with LTS, LMS, S, and MM. The data set is shown in Table 1.6, which contains 20 points and four of them (i=4, 6, 8, 19) are outliers (Rousseeuw, 1984). A linear regression model is used to investigate the influence of anatomical factors on wood specific gravity. The estimates of the six parameters by

OLS, LTS, S, MM, and LMS (LMS estimates are provided by Rousseeuw (1984)) are shown in Table 1. 7. LTS, S, and MM produce similar coefficient estimates and are close to LMS estimates. However, the OLS estimates are quite different from those robust estimates of LTS, S, MM, and LMS. Therefore, the OLS estimates are greatly affected by the outliers.

Table 1. 6 ***Modified Data on Wood Specific Gravity With Standardized Residuals From OLS, LTS, S, MM and LMS***

								Residual/Scale				
i	x_{i1}	x_{i2}	x_{i3}	x_{i4}	x_{i5}	x_{i6}	y_i	OLS	LTS	S	MM	LMS
1	0. 5730	0. 1059	0. 4650	0. 5380	0. 8410	1. 000	0. 5340	−0. 7250	2. 4231	0. 5728	0. 5937	−0. 0827
2	0. 6510	0. 1356	0. 5270	0. 5450	0. 8870	1. 000	0. 5350	0. 0472	0. 9400	0. 3391	0. 3306	0. 0013
3	0. 6060	0. 1273	0. 4940	0. 5210	0. 9200	1. 000	0. 5700	1. 2425	0. 0457	0. 2364	0. 0313	0. 2836
4	0. 4370	0. 1591	0. 4460	0. 4230	0. 9920	1. 000	0. 4500	0. 3546	−31. 878	−13. 599	−14. 067	−7. 6137
5	0. 5470	0. 1135	0. 5310	0. 5190	0. 9150	1. 000	0. 5480	1. 0023	2. 5153	1. 2121	0. 9750	0. 9020
6	0. 4440	0. 1628	0. 4290	0. 4110	0. 9840	1. 000	0. 4310	−0. 4518	−36. 752	−15. 813	−16. 268	−9. 1023
7	0. 4890	0. 1231	0. 5620	0. 4550	0. 8240	1. 000	0. 4810	0. 9066	2. 9574	0. 5036	0. 4830	0. 3746
8	0. 4130	0. 1673	0. 4180	0. 430	0. 9780	1. 000	0. 4230	−0. 0349	−35. 947	−15. 623	−15. 967	−8. 9077
9	0. 5360	0. 1182	0. 5920	0. 4640	0. 8540	1. 000	0. 4750	−0. 3959	0. 0457	−0. 4185	−0. 5688	−0. 3746
10	0. 6850	0. 1564	0. 6310	0. 5640	0. 9140	1. 000	0. 4860	−0. 4150	−0. 3174	−0. 0580	−0. 0245	−0. 2071
11	0. 6640	0. 1588	0. 5060	0. 4810	0. 8670	1. 000	0. 5540	1. 9856	0. 0457	−0. 2014	−0. 2003	0. 0013
12	0. 7030	0. 1335	0. 5190	0. 4840	0. 8120	1. 000	0. 5190	−1. 1975	0. 0457	−0. 5225	−0. 4747	−0. 9656
13	0. 6530	0. 1395	0. 6250	0. 5190	0. 8920	1. 000	0. 4920	−0. 4854	0. 8094	0. 3226	0. 2394	0. 0013
14	0. 5860	0. 1114	0. 5050	0. 5650	0. 8890	1. 000	0. 5170	−1. 2610	−0. 7948	−0. 4711	−0. 5230	−0. 6709
15	0. 5340	0. 1143	0. 5210	0. 5700	0. 8890	1. 000	0. 5020	−0. 5865	0. 6440	0. 0307	0. 0326	−0. 1733
16	0. 5230	0. 1320	0. 5050	0. 6120	0. 9190	1. 000	0. 5080	0. 5237	0. 0457	−0. 1238	−0. 0139	0. 0013
17	0. 5800	0. 1249	0. 5460	0. 6080	0. 9540	1. 000	0. 5200	−0. 2548	−0. 6450	0. 0344	−0. 0546	0. 0013
18	0. 4480	0. 1028	0. 5220	0. 5340	0. 9180	1. 000	0. 5060	0. 2838	−0. 9106	−0. 4673	−0. 6790	−0. 1090
19	0. 4170	0. 1687	0. 4050	0. 4150	0. 9810	1. 000	0. 4010	−1. 0836	−42. 291	−18. 381	−18. 770	−10. 726
20	0. 5280	0. 1057	0. 4240	0. 5660	0. 9090	1. 000	0. 5680	0. 5450	0. 0457	0. 0141	−0. 0990	0. 0013

Table 1.7 ***Regression Estimates for Modified Data on Wood Specific Gravity***

Estimators	β_1	β_2	β_3	β_4	β_5	Intercept
OLS	0.4407	-1.4750	-0.2612	0.0208	0.1708	0.4218
LTS	0.2384	-0.0699	-0.5695	-0.3839	0.6821	0.2880
S	0.2051	-0.1765	-0.5276	-0.4438	0.6163	0.3931
MM	0.2165	-0.0808	-0.5639	-0.3982	0.6046	0.3784
LMS	0.2687	-0.2381	-0.5357	-0.2937	0.4510	0.4347

Table 1.6 lists the standardized residuals (residuals divided by the estimated scale) for OLS, LTS, S, MM, and LMS (the standardized residuals for LMS are provided by Rousseeuw (1984)). The corresponding estimated scales are $\hat{\sigma}_{OLS}$ = 0.02412, $\hat{\sigma}_{LTS}$=0.0065, $\hat{\sigma}_S$=0.01351, $\hat{\sigma}_{MM}$=0.01351, and $\hat{\sigma}_{LMS}$=0. 0195, respectively. It is not easy to identify the outliers by looking at standardized residuals of OLS, but standardized residuals of LTS, S, MM and LMS could correctly identify the four outliers and the identified four outliers are exactly the same as which were spotted by LMS in Rousseeuw (1984). Therefore, the naive method by looking at the standardized residuals of OLS might miss the outliers due to the masking effect.

Figure 1.3 shows plots of the residuals versus the fitted values for OLS, LTS, S, MM and LMS, respectively. There are obvious four outliers by looking at residual plots of LTS, S, MM and LMS. However, the four outliers can not be easily detected by naively looking at the residual plot of OLS due to the masking effect.

Example 1.5: Finally, we apply OLS, LTS, LMS, S, and MM estimators to an artificial three-predictor data set which was created by Hawkins et al. (1984) to test the outlier detection of these regression parameter estimates. The dataset contains outliers in the cases 1-10. The standardized residuals for OLS, LTS, LMS, S, and MM

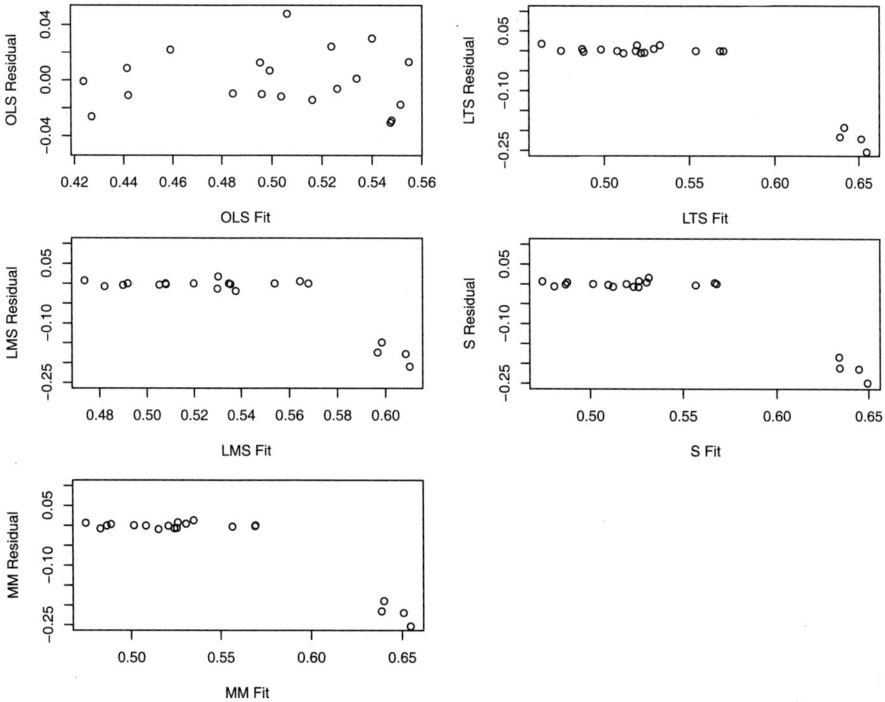

Figure 1.3 ***Plots of residual versus fitted values for OLS, LTS, S, MM and LMS for modified wood data***

estimations are given in Table 1. 8. The standardized residuals show that LTS, S, LMS, and MM all correctly flag the outliers and obtain similar coefficient estimation. Nonetheless, Hadi and Simonoff (1993) indicated that MM estimator with high efficiency level masked true outliers and swamped in the cases 11–14, but less-efficient versions of MM estimator (with efficiencies up to about 80%) give results similar to LMS and LTS. Similar to example 1.4, OLS estimator fails to identify the outliers.

Table 1.8 ***Hawkins, Bradu, and Kass Data With Standardized Residuals From OLS, LTS, LMS, S and MM***

i	OLS	LTS	LMS	S	MM	*i*	OLS	LTS	LMS	S	MM
1	1.4999	14.040	16.535	12.371	12.293	39	-0.5745	-0.6933	-0.6690	-0.6965	-0.7829
2	1.7759	14.847	17.406	13.032	12.854	40	-0.0099	-0.0116	-0.0556	-0.1934	-0.4685
3	1.3295	15.055	17.782	13.262	13.140	41	-0.4982	-0.1584	-0.0101	-0.1632	-0.1335
4	1.1369	14.155	16.687	12.433	12.195	42	-0.4855	0.0740	0.1245	-0.1518	-0.5418
5	1.3583	14.693	17.316	12.928	12.763	43	0.7704	1.5833	1.6742	1.1729	0.8701
6	1.5240	14.297	16.857	12.631	12.616	44	-0.8093	-0.3356	-0.2317	-0.4228	-0.6398
7	2.0050	15.516	18.205	13.672	13.623	45	-0.3459	-0.4275	-0.4002	-0.4718	-0.5562
8	1.7026	15.063	17.706	13.239	13.106	46	-0.6416	0.0743	0.2280	-0.0659	-0.2753
9	1.2022	14.338	16.897	12.582	12.337	47	-0.6773	-1.3703	-1.4135	-1.2232	-1.1638
10	1.3477	14.901	17.538	13.048	12.763	48	-0.2427	0.1680	0.3074	0.1052	0.1034
11	-3.4830	-0.1701	0.4452	0.0769	-0.0610	49	0.2863	1.1861	1.4335	0.9974	1.0416
12	-4.1743	-0.4320	0.3040	-0.1144	-0.2274	50	-0.3093	-0.2452	-0.2225	-0.3114	-0.4384
13	-2.7174	0.4759	1.1097	0.7177	0.7903	51	0.3901	1.0390	1.2209	0.8315	0.8249
14	-1.6666	-0.7451	-0.6690	-0.3591	-0.2921	52	-0.5249	-0.8739	-0.8425	-0.7766	-0.6907
15	-0.2979	-0.8288	-0.8548	-0.7288	-0.6373	53	-0.0263	2.2484	2.6351	1.7225	1.2866
16	0.3819	0.5360	0.6539	0.4493	0.5690	54	0.7640	1.3158	1.4109	1.0068	0.8646
17	0.2885	0.4903	0.4754	0.2189	-0.0845	55	0.3242	0.6418	0.6607	0.3775	0.1196
18	-0.1802	0.2414	0.3621	0.1251	0.0473	56	0.3487	0.2622	0.2572	0.1322	0.0642
19	0.2917	0.7430	0.7921	0.4788	0.2351	57	0.2719	1.3283	1.5702	1.0443	0.9420
20	0.1486	0.4199	0.5334	0.3356	0.3832	58	0.1222	-0.1701	-0.1946	-0.2107	-0.2211
21	0.2952	1.3877	1.6517	1.1399	1.1110	59	-0.3336	0.1523	0.2389	-0.0022	-0.2144
22	0.4179	1.1949	1.2841	0.8442	0.5344	60	-0.6017	-0.1735	-0.1587	-0.3697	-0.7796
23	-0.1919	-1.0632	-1.2060	-1.0068	-1.0502	61	-0.0071	0.4612	0.4772	0.1987	-0.1570
24	0.6022	1.3358	1.4797	1.0306	0.8950	62	0.3028	1.4909	1.7243	1.1312	0.9165
25	-0.1403	0.1607	0.2234	-0.0158	-0.2167	63	0.2911	-0.1379	-0.2334	-0.2523	-0.3870
26	-0.2130	-0.5121	-0.6183	-0.6089	-0.8698	64	-0.3994	-0.6549	-0.6636	-0.6225	-0.6498
27	-0.6181	-1.080	-1.1022	-0.9628	-0.9281	65	-0.1370	1.4073	1.7357	1.1044	0.9328
28	-0.1136	0.9259	1.1510	0.6831	0.5220	66	-0.1169	-0.5340	-0.6690	-0.6609	-0.9578

Table 1.8 ***Hawkins, Bradu, and Kass Data With Standardized Residuals From OLS, LTS, LMS, S and MM*** **(Continued table)**

i	OLS	LTS	LMS	S	MM	i	OLS	LTS	LMS	S	MM
29	0. 1722	0. 9506	1. 1064	0. 6807	0. 4976	67	−0. 2180	−0. 6910	−0. 7758	−0. 7146	−0. 8377
30	−0. 5705	0. 4733	0. 6579	0. 2399	−0. 0686	68	0. 2373	1. 8989	2. 1808	1. 4271	1. 0622
31	−0. 1257	−0. 1701	−0. 0921	−0. 1732	−0. 0960	69	0. 0814	0. 4751	0. 6002	0. 3214	0. 2680
32	0. 2492	−0. 0328	−0. 1266	−0. 2005	−0. 4262	70	0. 2082	1. 8047	2. 0823	1. 3933	1. 0981
33	−0. 0479	−0. 5604	−0. 6690	−0. 6050	−0. 7397	71	0. 0041	0. 5359	0. 6638	0. 3755	0. 2822
34	−0. 3046	−0. 1701	−0. 1718	−0. 3418	−0. 6366	72	0. 0611	0. 4370	0. 4695	0. 1981	−0. 0801
35	−0. 1840	0. 4784	0. 6690	0. 3729	0. 3664	73	0. 1951	1. 5509	1. 7583	1. 1355	0. 7729
36	−0. 5241	−0. 7504	−0. 8263	−0. 8108	−1. 0700	74	−0. 1719	−0. 3801	−0. 4942	−0. 5459	−0. 9029
37	−0. 0999	0. 1027	0. 0587	−0. 1279	−0. 5094	75	−0. 1588	1. 4297	1. 6891	1. 0198	0. 6185
38	0. 5546	1. 6095	1. 8378	1. 2772	1. 1646						

1.4 Discussion

In this chapter, we describe and compare different available robust methods. Table 1. 9 summarizes the robustness attributes and asymptotic efficiency of most of the estimators we have discussed. Based on Table 1. 11, it can be seen that MM-estimates and REWLSE have both high breakdown point and high efficiency. Our simulation study also demonstrated that MM-estimates and REWLSE have overall best performance among all compared robust methods. However, Park et al. (2012) pointed out that MM-estimates cannot detect any outliers when the contamination percentage is equal to and above 30%. In terms of breakdown point and efficiency, GM-estimates (Mallows, Schweppe), Bounded R-estimates, M-estimates, and LAD estimates are less attractive due to their low breakdown points. Although LMS, LTS, S - estimates, and GS - estimates are strongly resistent to outliers, their efficiencies are low. However, these high breakdown point robust estimates such as S-estimates and LTS are traditionally used as the initial estimates for some other high breakdown point and high efficiency robust estimates.

Table 1.9 ***Breakdown Points and Asymptotic Efficiencies of Various Regression Estimators***

	Estimator	Breakdown Point	Asymptotic Efficiency
High BP	LMS	0.5	0
	LTS	0.5	0.08
	S-estimates	0.5	0.29
	GS-estimates	0.5	0.67
	MM-estimates	0.5	0.95
	GM-estimates(S1S)	0.5	0.95
	REWLSE	0.5	1.00
Low BP	GM-estimates(Mallows, Schweppe)	$1/(p+1)$	0.95
	Bounded R-estimates	<0.2	0.90-0.95
	Monotone M-estimates	$1/n$	0.95
	LAD	$1/n$	0.64
	OLS	$1/n$	1.00

In theory, MM, S and LTS have high breakdown points. In practice, however, the com-putation of these estimators is very challenging (Hawkins and Olive, 2002; Park et al., 2012). One often uses the elemental resampling algorithm to obtain a number of subsets and calculates the initial regression estimate for each element set, and then computes the robust regression estimate from a number of initial estimates. In order to compute MM, S and LTS estimators with high breakdown point, one should consider all possible elemental sets because high breakdown property heavily depends on the number of elemental sets generated by resampling techniques. In addition, Hawkins and Olive (2002) proved that LTS estimator computed from the elemental resampling techniques, such as FAST-LTS algorithm, has zero breakdown point. Furthermore, Olive and Hawkins (2011) and Park et al. (2012) pointed out that if a practical initial estimator which has not been proved to be high breakdown is used in the implementation of the two-stage estimator such as S and MM estimator, the resulting two-stage estimator may be neither consistent nor high breakdown.

We would also like to mention some other directions to provide regression estimates which are robust to outliers. Lee (1989), Lee (1993), Kemp and Santos Silva (2012), and Yao and Li (2014) proposed *modal regression* to robustly estimate the regression function. Modal regression focuses on "most likely" conditional values rather than the conditional average or median. However, when the error distribution is homogeneous, modal regression line is the same as traditional mean regression line, except for intercepts. In addition, Linton and Xiao (2007), Yuan and De Gooijer (2007), Wang and Yao (2012), Yao and Zhao (2013), and Chen et al. (2015) proposed to adaptively estimate the regression functions by estimating the error density using kernel density estimation. Those adaptive estimates are also demonstrated to be robust to outliers and heavy-tail error distributions based on their simulation studies.

Chapter 2
A Selective Overview and Comparison of Robust Mixture Regression Estimators

2.1 Introduction

Finite mixture regression (FMR) model provides a flexible statistical tool in studying the in-homogeneous population which contains several homogeneous sub-populations (components) and the relationship between the response y and the predictors $\boldsymbol{x}$ varies across the sub-populations. Since the FMR model was first introduced by Goldfeld and Quandt (1973), it has been widely used in many areas such as business, marketing, social sciences, etc; see in literature, e.g., Jiang and Tanner (1999), Böhning (1999), Wedel (2000), Henning (2000), McLachlan and Peel (2000), Skrondal and Rabe-Hesketh (2004), and Frühwirth-Schnatter (2006). The FMR model assumes that the response y depends on the explanatory variable $\boldsymbol{x}$ in a linear way for each component as $y = \boldsymbol{x}^{\mathrm{T}}\beta_j + \epsilon_j$, where $(y, \boldsymbol{x})$ belongs to the jth component $(j = 1, 2, \cdots, m)$, $\beta_j \in$

$\mathbb{R}^p$ is a fixed and unknown coefficient vector, and $\epsilon_j \sim N(0,\sigma_j^2)$ with $\sigma_j^2>0$. Note that the intercept term can be included by setting the first element of each $\boldsymbol{x}$ vector as one. Then the conditional density of y given $\boldsymbol{x}$, is

$$f(y|\boldsymbol{x},\boldsymbol{\theta})=\sum_{j=1}^{m}\pi_j\phi(y;\boldsymbol{x}^{\mathrm{T}}\boldsymbol{\beta}_j,\sigma_j^2) \tag{2-1}$$

where $\phi(\cdot;\mu,\sigma^2)$ denotes the probability density function (pdf) of the normal distribution $N(\mu,\sigma^2)$, π_j's are the mixing proportions with $\sum_{j=1}^{m}\pi_j=1$, and $\boldsymbol{\theta}=(\pi_1,\boldsymbol{\beta}_1^{\mathrm{T}},\sigma_1;\dots;\pi_m,\boldsymbol{\beta}_m^{\mathrm{T}},\sigma_m)^{\mathrm{T}}$ is the unknown parameter vector.

Let $\{(\mathbf{x}_i,y_i),i=1,\dots,n\}$ be the collected observations from the mixture density of (2-1). Maximum likelihood estimation (MLE) is commonly used to estimate $\boldsymbol{\theta}$ in (2-1) by maximizing the following log-likelihood function

$$\ell_n(\boldsymbol{\theta})=\sum_{i=1}^{n}\log\left\{\sum_{j=1}^{m}\pi_j\phi(y_i;\boldsymbol{x}_i^{\mathrm{T}}\boldsymbol{\beta}_j,\sigma_j^2)\right\} \tag{2-2}$$

The MLE does not have an explicit form in general and is usually obtained by the Expectation-Maximization (EM) algorithm (Dempster et al. 1977).

Let

$$z_{ij}=\begin{cases}1 & \text{if the } i\text{th observation is from the } j\text{th component}\\ 0 & \text{otherwise}\end{cases}$$

Denote the complete data set by $\{(\boldsymbol{x}_i,z_i,y_i):i=1,2,\dots,n\}$, where the component labels $\boldsymbol{z}_i=(z_{i1},z_{i2},\dots,z_{im})$ are not observable. Then, the complete log-likelihood function is

$$\ell_n^c(\boldsymbol{\theta})=\sum_{i=1}^{n}\sum_{j=1}^{m}z_{ij}\log\{\pi_j\phi(y_i-\boldsymbol{x}_i^{\mathrm{T}}\boldsymbol{\beta}_j;0,\sigma_j^2)\} \tag{2-3}$$

$$= \sum_{i=1}^{n}\sum_{j=1}^{m} z_{ij}\left\{\log(\pi_j) - \frac{1}{2}\log(2\pi) - \frac{1}{2}\log(\sigma_j^2)\right\} - \sum_{i=1}^{n}\sum_{j=1}^{m} z_{ij}\frac{(y_i - \boldsymbol{x}_i^{\mathrm{T}}\boldsymbol{\beta}_j)^2}{2\sigma_j^2} \tag{2-4}$$

The E step of EM algorithm computes the conditional expectation of the complete log-likelihood function $E[\ell_n^c(\boldsymbol{\theta})|\mathbf{y}, \mathbf{X}]$ and the M step updates $\boldsymbol{\theta}$ by maximizing $E[\ell_n^c(\boldsymbol{\theta})|\mathbf{y}, \mathbf{X}]$, where $\mathbf{y}=(y_1, \dots, y_n)$ and $\mathbf{X}=(\mathrm{x}_1, \dots, \mathrm{x}_n)^{\mathrm{T}}$. Specifically, the EM algorithm for the model (2.1) can be summarized as below:

Algorithm 2.1: *Given the initial parameter estimate* $\boldsymbol{\theta}^{(0)}$, at $(k+1)$*th step, the EM algorithm iterates the following E-step and M-step*:

E-step: *Calculate the classification probabilities*:

$$p_{ij}^{(k+1)} = \frac{\pi_j^{(k)}\phi(y_i - \boldsymbol{x}_i^{\mathrm{T}}\boldsymbol{\beta}_j^{(k)}; 0, \sigma_j^{2(k)})}{\sum_{j=1}^{m}\pi_j^{(k)}\phi(y_i - \boldsymbol{x}_i^{\mathrm{T}}\boldsymbol{\beta}_j^{(k)}; 0, \sigma_j^{2(k)})} \qquad i = 1, \dots, n; j = 1, \dots, m \tag{2-5}$$

M-step: *Update parameter estimates*:

$$\pi_j^{(k+1)} = (\sum_{i=1}^{n} p_{ij}^{(k+1)})/n$$

$$\boldsymbol{\beta}_j^{(k+1)} = \arg\min_{\boldsymbol{\beta}_j}\sum_{i=1}^{n} p_{ij}^{(k+1)}(y_i - \boldsymbol{x}_i^{\mathrm{T}}\boldsymbol{\beta}_j)^2$$

$$= \left(\sum_{i=1}^{n} p_{ij}^{(k+1)}\boldsymbol{x}_i\boldsymbol{x}_i^{\mathrm{T}}\right)^{-1}\left(\sum_{i=1}^{n} p_{ij}^{(k+1)}\boldsymbol{x}_i y_i\right)$$

$$\sigma_j^{2(k+1)} = \frac{\sum_{i=1}^{n}\sum_{j=1}^{m} p_{ij}^{(k+1)}(y_i - \boldsymbol{x}_i^{\mathrm{T}}\boldsymbol{\beta}_j^{(k+1)})^2}{\sum_{i=1}^{n} p_{ij}^{(k+1)}}$$

where $j=1, \dots, m$.

It is well known that the log-likelihood function (2.2) is unbounded and goes to infinity if σ_j parameters are close to zero. This type of degeneracy creates troubles in the EM algorithm and results in the detection of non-interesting solutions. To circumvent this problem, Hathaway (1985, 1986) proposed putting some constraints on the parameter space such that the component variance has some low limit. We restrict $(\sigma_1, \ldots, \sigma_m) \in \Omega_\sigma$, with Ω_σ defined as

$$\Omega_\sigma = \{(\sigma_1, \ldots, \sigma_m) : \sigma_j > 0, \text{ for } 1 \leqslant j \leqslant m$$
$$\text{and } \sigma_j / \sigma_k \geqslant \epsilon, \text{ for } j \neq k \text{ and } 1 \leqslant j, k \leqslant m\}$$

where ϵ is a very small positive value. In the examples that follow, we set $\epsilon =$ 0.01. Accordingly, we define the parameter space of $\boldsymbol{\theta}$ as

$$\Omega = \{(\pi_j, \boldsymbol{\beta}_j, \sigma_j), j = 1, \ldots, m : 0 \leqslant \pi_j \leqslant 1$$
$$\sum_{j=1}^{m} \pi_j = 1, (\sigma_1, \ldots, \sigma_m) \in \Omega_\sigma\}$$

The MLE can be very sensitive to outliers in a data set. This can be partly seen by the least squares update of regression parameter $\boldsymbol{\beta}_j$ in the above EM algorithm (3.2.1). In fact, even one single atypical value may have a large effect on the parameter estimates. To over-come this problem, many robust methods for mixture regression models have been developed recently. One class of robust mixture regression methods was proposed to replace the normal error densities by some heavy-tailed error densities. For example, Yao et al. (2014), Zhang (2013), Zeller et al. (2016), and Song et al. (2014) considered robust mixture regression using t-distribution, Pearson type Ⅶ distribution, scale mixtures of skew-normal distributions, and laplace distribution, respectively. García-Escudero et al. (2017) proposed another new robust estimator for mixtures of regression based on the trimmed cluster weighted restricted model (CWRM). Punzo and McNicholas (2017) proposed robust estimation of mixture regression via the contaminated gaussian cluster weighted model (CWM). Neykov et al. (2007) proposed to fit the mixture model using the trimmed likelihood method. Doğru and Arslan (2017) proposed adapted complete data log-likelihood function using the least trimmed squares (LTS) criterion. Bai et al.(2012) developed a modified EM algorithm by adopting a robust criterion in the M-step. Hu et al. (2017) proposed a robust EM-type algorithm by assuming the component error densities are log-concave. Markatou (2000) and Shen et al. (2004) proposed to weight each data point in order to robustify the estimation procedure. Bashir and Carter (2012) extended the idea of the S-estimator to mixture regression.

In Chapter 2, we give a selective overview of some recently developed robust methods for mixture regression models and use a simulation study to compare the performance of some of them. We will review and describe some of the available robust methods in more detail in Section 2.2. In Section 2.3, simulation studies are used to compare different robust methods. The conclusion and some discussions are given in Section 2.4.

2.2 Robust mixture regression methods

Many extensions of the classic Gaussian mixture regression model have been proposed by replacing the normal error densities by some heavy-tailed error densities. In Section 2.2.1-2.2.4, we introduce some of those developed methods.

2.2.1 Robust mixture regression using the t-distribution

Yao et al. (2014) proposed a robust mixture regression method using a t-distribution. The density function of y_i given x_i is

$$f(y_i; \mathbf{x}_i, \boldsymbol{\theta}) = \sum_{j=1}^{m} \pi_j g_t(y_i - \boldsymbol{x}_i^{\mathrm{T}} \boldsymbol{\beta}_j; \sigma_j^2, \nu_j)$$

where the component error density $g_t(y_i - \boldsymbol{x}_i^T \boldsymbol{\beta}_j; \sigma_j^2, \nu_j)$ is a t-distribution with degree of freedom ν and scale parameter σ

$$g_t(y_i - \boldsymbol{x}_i^{\mathrm{T}} \boldsymbol{\beta}_j; \sigma_j^2, \nu_j) = \frac{\Gamma(\frac{\nu+1}{2})\sigma^{-1}}{(\pi\nu)^{\frac{1}{2}}\Gamma(\frac{\nu}{2})\left\{\frac{\Delta^2(y_i - \boldsymbol{x}_i^{\mathrm{T}}\boldsymbol{\beta}_j; \sigma_j^2)}{\nu}\right\}^{\frac{1}{2}(\nu+1)}}$$

and

$$\Delta^2(y_i - \boldsymbol{x}_i^{\mathrm{T}} \boldsymbol{\beta}_j; \sigma_j^2) = (y_i - \boldsymbol{x}_i^{\mathrm{T}} \boldsymbol{\beta}_j)^2 / \sigma_j^2 \tag{2-6}$$

Note that the normal density is a special case of t-distribution if $\nu \to \infty$. Therefore, by adaptively choose ν Yao et al. (2014) argued their new robust estimator also has the full efficiency compared to the traditional normality based MLE. The unknown parameter $\boldsymbol{\theta}$ can be estimated by maximizing the log-likelihood

$$\ell(\boldsymbol{\theta}) = \sum_{i=1}^{n} \log \left\{ \sum_{j=1}^{m} \pi_j g_t(y_i - \boldsymbol{x}_i^{\mathrm{T}} \boldsymbol{\beta}_j; \sigma_j, \nu_j) \right\} \tag{2-7}$$

An EM algorithm was proposed by Yao et al. (2014) to maximize (2-7). The computation of EM algorithm can be simplified by introducing another latent variable u. Let u be the latent variable such that

$$\epsilon|u \sim N(0,\sigma^2/u),\ u \sim \text{gamma}(\frac{1}{2}\nu, \frac{1}{2}\nu)$$

where gamma(α, γ) has density

$$f(u;\alpha,\gamma) = \frac{1}{\Gamma(\alpha)}\gamma^{\alpha}u^{\alpha-1}e^{-\gamma u},\ u>0 \tag{2-8}$$

Then, marginally ϵ has a t-distribution with degree of freedom ν and scale parameter σ.

Algorithm 2.2: *Given the initial parameter estimate* $\boldsymbol{\theta}^{(0)}$, *at* $(k+1)$*th step the EM algorithm iterates the following E-step and M-step*

E-step: *Calculate the classification probabilities*:

$$p_{ij}^{(k+1)} = E\left(z_{ij} \mid \boldsymbol{X}, \boldsymbol{y}, \boldsymbol{\theta}^{(k)}\right) = \frac{\pi_j^{(k)} g_t(y_i - \boldsymbol{x}_i^{\mathrm{T}}\beta_j^{(k)}; \sigma_j^{(k)}, \nu_j)}{\sum_{j=1}^m \pi_j^{(k)} g_t(y_i - \boldsymbol{x}_i^{\mathrm{T}}\beta_j^{(k)}; \sigma_j^{(k)}, \nu_j)}$$

and

$$u_{ij}^{(k+1)} = E(u_i \mid \boldsymbol{X}, \boldsymbol{y}, \boldsymbol{\theta}^{(k)}, z_{ij}=1) = \frac{\nu+1}{\nu + \Delta^2(y_i - \boldsymbol{x}_i^{\mathrm{T}}\boldsymbol{\beta}_j^{(k)}; \sigma_j^{2(k)})}$$

M-step: *Update parameter estimates*:

$$\pi_j^{(k+1)} = \sum_{i=1}^n p_{ij}^{(k+1)}/n$$

$$\boldsymbol{\beta}_j^{(k+1)} = \left(\sum_{i=1}^n \boldsymbol{x}_i\boldsymbol{x}_j^{\mathrm{T}} p_{ij}^{(k+1)} u_{ij}^{(k+1)}\right)^{-1} \left(\sum_{i=1}^n \boldsymbol{x}_i y_i p_{ij}^{(k+1)} u_{ij}^{(k+1)}\right)$$

$$\sigma_j^{2(k+1)} = \frac{\sum_{i=1}^n \sum_{j=1}^m p_{ij}^{(k+1)} u_{ij}^{(k+1)} (y_i - \boldsymbol{x}_i^{\mathrm{T}}\boldsymbol{\beta}_j^{(k+1)})^2}{\sum_{i=1}^n p_{ij}^{(k+1)}}$$

where $j=1, \ldots, m$.

In addition, Yao et al. (2014) proposed to use the profile likelihood to adaptively choose the degrees of freedom for the t-distribution. Given the kth estimate

$\boldsymbol{\theta}^{(k)}$, the degrees of freedom ν_j can be updated at the ($k+1$)th iteration of M step by

$$\nu_j^{(k+1)} = \arg\max_{\nu_j} \sum_{i=1}^{n} p_{ij}^{(k+1)} \Big[-\log\Gamma(0.5\nu_j) + 0.5\nu_j \log(0.5\nu_j) + 0.5\nu_j \left\{ v_{ij}^{(k+1)} - u_{ij}^{(k+1)} \right\} - v_{ij}^{(k+1)} \Big]$$

The proposed estimate has high efficiency, i.e., comparable performance to the traditional MLE when the errors are normally distributed, due to the adaptive choice of degrees of freedom.

Similar to the traditional M-estimate for linear regression (Maronna et al., 2006), the proposed mixture of regression based on t-distribution is sensitive to the high leverage outliers. To overcome this problem, Yao et al. (2014) proposed a trimmed version of the new method by fitting the new model to the data after adaptively trimming the high leverage points.

Let $\mathbf{X} = (\mathbf{x}_1, \ldots, \mathbf{x}_n)^T$ and h_{ii} be the ith diagonal of H, where $H = \mathbf{X}(\mathbf{X}^T\mathbf{X})^{-1}\mathbf{X}^T$. Then, h_{ii} is called the leverage for ith predictor $\boldsymbol{x}_i$ and $\boldsymbol{x}_i$ is considered as a high leverage point if h_{ii} is large.

Note however

$$h_{ii} = n^{-1} + (n-1)^{-1}\mathrm{MD}_i$$

where $M D_i$ is defined as a modified Mahalanobis distance

$$\mathrm{MD}_i = (\mathbf{x}_i - \mathbf{m}(\mathbf{X}))^T \mathbf{C}(\mathbf{X})^{-1} (\mathbf{x}_i - \mathbf{m}(\mathbf{X}))$$

where $\mathbf{m}(\mathbf{X})$ and $\mathbf{C}(\mathbf{X})$ are robust estimates of location and scatter for $\mathbf{X}$ (after removing the first column 1s).

Yao et al. (2014) proposed to use the minimum covariance determinant

(MCD, Rosseeuw, 1984) estimators for $\mathbf{m}(\mathbf{X})$ and $\mathbf{C}(\mathbf{X})$ and implement it by Fast MCD algorithm of Rousseeuw and Van Driessen (1999). After obtaining the robust estimate MD_i, Yao et al. (2014) proposed to trim the observations with $MD_i > \chi^2_{p-1,0.975}$ and thus to make the proposed method robust against the high leverage outliers.

2.2.2 Robust mixture regression modeling using Pearson type Ⅶ distribution

Inspired by the research work of Sun et al. (2010), which developed a robust mixture clustering using Pearson type Ⅶ distribution, Zhang (2013) extended the robust method to FMR models. The Pearson type Ⅶ distribution is defined as follows

$$g_d(\mathbf{y};\boldsymbol{\mu},\boldsymbol{\Sigma},\nu)=\frac{\Gamma(\nu)}{\pi^{\frac{d}{2}}\Gamma(\nu-\frac{d}{2})}|\boldsymbol{\Sigma}|^{-\frac{1}{2}}(1+\Delta^2)^{-\nu}$$

where $\Delta^2=(\mathrm{y}-\boldsymbol{\mu})^{\mathrm{T}}\Sigma^{-1}(\mathbf{y}-\boldsymbol{\mu})$ is the Mahalanobis distance and $2\nu>d$, ν is the degree of freedom that controls the degree of robustness, and d is the dimensionality of $\mathbf{y}$. It can be shown that the student t-distribution $S(\mathbf{y};\boldsymbol{\mu}\ \Sigma,v)$ with degree of freedom v

$$S(\mathbf{y};\boldsymbol{\mu},\boldsymbol{\Sigma},v)=\frac{\Gamma(\frac{v+d}{2})|\boldsymbol{\Sigma}|^{-\frac{1}{2}}}{(\pi v)^{\frac{d}{2}}\Gamma(\frac{v}{2})(1+\frac{\Delta^2}{v})^{\frac{v+d}{2}}}$$

is a special case of the Pearson type Ⅶ distribution if we set $\nu=(v+d)/2$ and $\Sigma=\Sigma v$.

In order to robustly estimate the mixture regression parameters, Zhang (2013) proposed the following density function of y_i given $\boldsymbol{x}_i$

$$f(y_i;\boldsymbol{x}_i,\boldsymbol{\theta})=\sum_{j=1}^{m}\pi_j g_1(y_i-\boldsymbol{x}_i^{\mathrm{T}}\boldsymbol{\beta}_j;\sigma_j^2,\nu_j) \tag{2-9}$$

where

$$g_1(y_i-\boldsymbol{x}_i^{\mathrm{T}}\boldsymbol{\beta}_j;\sigma_j^2,\nu_j)=\frac{\Gamma(\nu_j)\sigma_j^{-1}}{\pi_j^{\frac{1}{2}}\Gamma(\nu_j-\frac{1}{2})\left\{1+\Delta^2(y_i-\boldsymbol{x}_i^{\mathrm{T}}\boldsymbol{\beta}_j;\sigma_j^2)\right\}^{\nu_j}}$$

and $\Delta^2(y_i-\mathbf{x}_i^{\mathrm{T}}\beta_j;\ \sigma_j^2)$ is defined in the same way as in (2-6).

The unknown parameter $\boldsymbol{\theta}$ can be estimated by maximizing the log-likelihood

$$\ell(\boldsymbol{\theta})=\sum_{i=1}^{n}\log\left\{\sum_{j=1}^{m}\pi_j g_1(y_i-\boldsymbol{x}_i^{\mathrm{T}}\boldsymbol{\beta}_j;\sigma_j^2,\nu_j)\right\}$$

Since the Pearson type Ⅶ distribution can be considered as a scale mixture of normal distribution, another latent variable u is introduced. Let u be the latent variable such that

$$\epsilon|u\sim N(\boldsymbol{x}^{\mathrm{T}}\boldsymbol{\beta},\sigma^2/u),\ u\sim\mathrm{gamma}(\nu-\frac{1}{2},\frac{1}{2})$$

where the density of gamma(α, $\boldsymbol{\beta}$) is defined in (2-8). Then, marginally ε has a Pearson type Ⅶ distribution with degree of freedom ν and scale parameter $\boldsymbol{\sigma}$.

Zhang (2013) proposed the following EM algorithm to estimate the model (2-9).

Algorithm 2.3: *Given the initial parameter estimate* $\boldsymbol{\theta}^{(0)}$, *at* $(k+1)$*th step, the EM algorithm iterates the following E-step and M-step*

E-step: *Calculate the classification probabilities*:

$$p_{ij}^{(k+1)}=\frac{\pi_j^{(k)}g_1(y_i-\boldsymbol{x}_i^{\mathrm{T}}\beta_j^{(k)};\sigma_j^{2(k)},\nu_j^{(k)})}{\sum_{j=1}^{m}\pi_j^{(k)}g_1(y_i-\boldsymbol{x}_i^{\mathrm{T}}\beta_j^{(k)};\sigma_j^{2(k)},\nu_j^{(k)})}$$

and

$$u_{ij}^{(k+1)} = \frac{2\nu_j^{(k)}}{1+\Delta^2(y_i - \boldsymbol{x}_i^{\mathrm{T}}\beta_j^{(k)}; \sigma_j^{2(k)})},$$

where $i=1, \ldots, n, j=1, \ldots, m.$

M-step: *Update parameters* $\pi_j^{(k+1)}$, $\boldsymbol{\beta}_j^{(k+1)}$ *and* $\sigma_j^{2(k+1)}$ *using the same formulas as in the M step of EM algorithm* (2-2) *in Section* 2.2.1.

Furthermore, given the kth estimate $\boldsymbol{\theta}^{(k)}$, the degrees of freedom ν_j can be updated at the $(k+1)$th iteration of M step. To simplify the computation, assume that $\nu_1 = \nu_2 = \ldots = \nu_m = \nu$ and the estimation of ν can be obtained by

$$\hat{\nu}^{(k+1)} = \arg\max_{\nu} \sum_{i=1}^{n} \log\left\{\sum_{j=1}^{m} \pi_j^{(k+1)} g_1(y_i - \boldsymbol{x}_i^{\mathrm{T}}\boldsymbol{\beta}_j^{(k+1)}; \sigma_j^{2(k+1)}, \nu)\right\} \tag{2-10}$$

In the above function (2-10), ν can be substituted simply by a set of values: $\nu = 0.6, \ldots, \nu_{\max}$. Usually, the value of $\nu_{\max}$ does not need to be too large and 15 to 20 would be enough.

This mixture regression model (2-9) is robust when the outliers are in y-direction, but not in x-direction. To make the proposed method also robust against the high leverage outliers, a similar modified version of new method to Yao et al. (2014) is required by fitting the new model to the data after adaptively trimming the high leverage points.

2.2.3 Robust mixture regression model fitting by Laplace distribution

Due to the robustness of least absolute deviation (LAD) procedure, Song et al. (2014) proposed a robust mixture regression model fitting by Laplace distribution. Andrews and Mal-lows (1974) showed that a Laplace distribution can be expressed as a mixture of a normal distribution and another distribution related to the exponential distribution. To be specific, suppose Z and V are two random variables, V has a distribution with density function $v^{-3}\exp(-(2v^2)^{-1})$, $v>0$, and given $V=v$ the conditional distribution of Z is normal with mean 0 and variance $\sigma^2/(2v^2)$. Then Z marginally has a Laplace distribution with density function $h_\varepsilon(z)=\exp(-\sqrt{2}\,|z|/\sigma)/(\sqrt{2}\,\sigma)$. Song et al. (2014) proposed to model the component error terms by a Laplace distribution with mean 0 and scale parameter $1/\sqrt{2}$. The unknown parameter $\boldsymbol{\theta}$ can be estimated by maximizing the log-likelihood

$$\ell(\boldsymbol{\theta})=\sum_{i=1}^{n}\log\left\{\sum_{j=1}^{m}\frac{\pi_j}{\sqrt{2}\sigma_j}\exp\left(-\frac{\sqrt{2}|y_i-\boldsymbol{x}_i^{\mathrm{T}}\boldsymbol{\beta}_j|}{\sigma_j}\right)\right\} \tag{2-11}$$

The following EM algorithm was proposed by Song et al. (2014) to maximize (2-11).

Algorithm 2.4: *Given the initial parameter estimate $\boldsymbol{\theta}^{(0)}$, we iterate the following E-step and M-step.*

E-step: *Calculate*

$$p_{ij}^{(k+1)}=\frac{\pi_j^{(k)}\sigma_j^{-1(k)}\exp(-\sqrt{2}|y_i-\boldsymbol{x}_i^{\mathrm{T}}\beta_j^{(k)}|/\sigma_j^{(k)})}{\sum_{j=1}^{m}\pi_j^{(k)}\sigma_j^{-1(k)}\exp(-\sqrt{2}|y_i-\boldsymbol{x}_i^{\mathrm{T}}\beta_j^{(k)}|/\sigma_j^{(k)})}$$

and

$$u_{ij}^{(k+1)}=\frac{\sigma_j^{(k)}}{\sqrt{2}|y_i-\boldsymbol{x}_i^{\mathrm{T}}\beta_j^{(k)}|}$$

M-step: *Update parameters* $\pi_j^{(k+1)}$, $\beta_j^{(k+1)}$ *and* $\sigma_j^{2(k+1)}$ *using the same formulas as in the M step of EM algorithm* (2.2) *in Section* 2.2.1.

Note that the t-distribution based procedure (Yao et al. 2014) achieves both high efficiency and robustness by requiring a data adaptive choice/tuning of the degree of freedom. In comparison, the Laplace distribution based procedure has simpler and faster computation without choosing any tuning parameter, but with the loss of some efficiency when there are no outliers in the data.

2.2.4 Robust mixture regression modeling based on Scale mixtures of skew-normal distributions

Zeller et al. (2016) proposed a robust mixture regression modeling based on scale mixtures of skew-normal (SMSN) distributions by generalizing the recent works of Basso et al. (2010) and Yao et al. (2014). The proposed robust mixture regression model can simultaneously accommodate asymmetry and heavy tails.

Definition 2.1: A random variable Y is said to have a skew-normal distribution with a location parameter μ, dispersion parameter $\sigma^2>0$ and skewness parameter $\boldsymbol{\lambda}$, and we write $Y \sim SN(\mu, \sigma^2, \boldsymbol{\lambda})$, if its density is given by

$$f(y) = 2\phi(y; \mu, \sigma^2)\Phi(a), \qquad y \in \mathbb{R}$$

where $a=\lambda\sigma^{-1}(y-\mu)$, $\phi(\cdot; \mu, \sigma^2)$ stands for the pdf of the univariate normal distribution with mean μ and variance σ^2, and $\Phi(\cdot)$ represents the distribution function of the standard univariate normal distribution.

Definition 2.2: The distribution of the random variable Y belongs to the family of SMSN distributions when

$$Y = \mu + K(U)^{1/2}X$$

where μ is a location parameter, $X \sim SN(0, \sigma^2, \lambda)$, $K(\cdot)$ is a positive weight function and U is a positive random variable with a cdf $H(u; \boldsymbol{\nu})$, where $\boldsymbol{\nu}$ is a parameter indexing the distribution of U, known as the scale factor parameter, which is independent of X.

The SMSN distribution is written as $Y \sim SMSN(\mu, \sigma^2, \lambda; H)$. The name of the class becomes clear when we note that the conditional distribution of Y given $U=u$ is skew-normal. Specifically, we have that

$$Y|U=u \sim SN(\mu, K(u)\sigma^2, \lambda), \qquad U \sim H(.;\boldsymbol{\nu})$$

Thus, the density of Y is given by

$$g(y) = 2\int_0^\infty \phi(y;\mu, K(u)\sigma^2)\Phi(K(u)^{-1/2}a)dH(u;\boldsymbol{\nu})$$

The SMSN distribution is also written as $Y \sim SMSN(\mu, \sigma^2, \lambda, \boldsymbol{\nu})$.

The normal mixture of regression models are extended by considering the following error assumption

$$\epsilon_j \sim SMSN(b\Delta_j, \sigma_j^2, \lambda_j, \boldsymbol{\nu}_j), \qquad j=1,\dots,m$$

where $\Delta_j = \sigma_j \delta_j$, $\delta_j = \frac{\lambda_j}{\sqrt{1+\lambda_j^2}}$, $b = -\sqrt{\frac{2}{\pi}}K_1$, with $K_r = E[K^{r/2}(U)]$, $r=1, 2, \ldots$, which corresponds to the regression model where the error distribution has mean zero and hence the regression parameters are all comparable. Thus the mixture regression modeling based on SMSN distributions proposed by Zeller et al. (2016) is defined as follows:

$$f(y;\boldsymbol{x},\boldsymbol{\theta}) = \sum_{j=1}^m \pi_j g(y;\boldsymbol{x},\boldsymbol{\theta}_j) \tag{2-12}$$

where $g(\cdot; \boldsymbol{x}, \boldsymbol{\theta}_j)$ is the density function of $SMSN(\boldsymbol{x}^T\beta_j + b\Delta_j, \sigma_j^2, \lambda_j, \nu_j)$

and $\boldsymbol{\theta}_j=(\pi_j, \beta_j, \sigma_j^2, \lambda_j, \boldsymbol{\nu}_j)^{\mathrm{T}}$. For computational convenience, it is assumed that $\boldsymbol{\nu}_1=\boldsymbol{\nu}_2=\ldots=\boldsymbol{\nu}_m=\boldsymbol{\nu}$.

Zeller et al. (2016) proposed the following EM-type algorithm (ECME) to estimate the parameters in (2-12):

Algorithm 2.5: *Given the initial value* $\hat{\boldsymbol{\theta}}^{(0)}$, *at* $(k+1)$*th step*,

E-step : *compute* $\widehat{z}_{ij}^{(k)}$, $\widehat{zu}_{ij}^{(k)}$, $\widehat{zut}_{ij}^{(k)}$ *and* $\widehat{zut2}_{ij}^{(k)}$ *for* $i=1, \ldots, n$, *using the following formulas*:

$$\widehat{z}_{ij}^{(k)}=\frac{\pi_j^{(k)}g(y_i;\boldsymbol{x}_i,\widehat{\boldsymbol{\theta}}_j^{(k)})}{\sum_{j=1}^m\pi_j^{(k)}g(y_i;\boldsymbol{x}_i,\widehat{\boldsymbol{\theta}}_j^{(k)})}$$

$$\widehat{zu}_{ij}^{(k)}=\widehat{z}_{ij}^{(k)}\widehat{u}_{ij}^{(k)}$$

$$\widehat{zut}_{ij}^{(k)}=\widehat{z}_{ij}^{(k)}\widehat{ut}_{ij}^{(k)}$$

$$\widehat{zut2}_{ij}^{(k)}=\widehat{z}_{ij}^{(k)}\widehat{ut2}_{ij}^{(k)}$$

where $\widehat{ut}_{ij}^{(k)}=\widehat{u}_{ij}^{(k)}(\widehat{m}_{ij}^{(k)}+b)+\widehat{M}_j^{(k)}\widehat{\eta}_{ij}^{(k)}$, $\widehat{ut2}_{ij}^{(k)}=\widehat{u}_{ij}^{(k)}(\widehat{m}_{ij}^{(k)}+b)^2+\widehat{M}_j^{2(k)}+\widehat{M}_j^{(k)}(\widehat{m}_{ij}^{(k)}+2b)\widehat{\eta}_{ij}^{(k)}$, *with* $\widehat{M}_j^2=\frac{\widehat{\gamma}_j^2}{\widehat{\gamma}_j^2+\widehat{\Delta}_j^2}$ *and* $\widehat{m}_{ij}=\widehat{M}_j^2\frac{\widehat{\Delta}_j}{\widehat{\gamma}_j^2}(y_i-\boldsymbol{x}_i^{\mathrm{T}}\widehat{\boldsymbol{\beta}}_j-b\widehat{\Delta}_j)$, $i=1,\ldots,n$, *evaluated at* $\boldsymbol{\theta}=\widehat{\boldsymbol{\theta}}^{(k)}$.

CM-step : *update* $\hat{\boldsymbol{\theta}}^{(k+1)}$ *using the following closed form expressions*:

$$\pi_j^{(k+1)}=(\sum_{i=1}^n\widehat{z}_{ij}^{(k)})/n$$

$$\boldsymbol{\beta}_j^{(k+1)}=\left(\sum_{i=1}^n\widehat{zu}_{ij}^{(k)}\boldsymbol{x}_i\boldsymbol{x}_i^{\mathrm{T}}\right)^{-1}\sum_{i=1}^n\left(\widehat{zu}_i^{(k)}y_i-\widehat{zut}_{ij}^{(k)}\widehat{\Delta}_j^{(k)}\right)\boldsymbol{x}_i$$

$$\widehat{\gamma}_j^{2(k+1)}=\frac{\sum_{i=1}^n\left[\widehat{zu}_{ij}^{(k)}(y_i-\boldsymbol{x}_i^{\mathrm{T}}\boldsymbol{\beta}_j^{(k)})^2-2\widehat{zut}_{ij}^{(k)}\widehat{\Delta}_j^{(k)}(y_i-\boldsymbol{x}_i^{\mathrm{T}}\boldsymbol{\beta}_j^{(k)})+\widehat{zut2}_{ij}^{(k)}\widehat{\Delta}_j^{2(k)}\right]}{\sum_{i=1}^n\widehat{z}_{ij}^{(k)}}$$

$$\widehat{\Delta}_j^{(k+1)} = \frac{\widehat{zut}_{ij}^{(k)}(y_i - \boldsymbol{x}_i^{\mathrm{T}}\boldsymbol{\beta}_j^{(k)})}{\sum_{i=1}^{n}\widehat{zut^2}_{ij}^{(k)}}$$

$$\sigma_j^{2(k+1)} = \widehat{\gamma}_j^{2(k+1)} + \widehat{\Delta}_j^{2(k+1)}$$

$$\widehat{\lambda}_j^{(k+1)} = \frac{\widehat{\Delta}_j^{(k+1)}}{\sqrt{\widehat{\gamma}_j^{2(k+1)}}}, \qquad j = 1, \ldots, m$$

***CML-step**: Update $\hat{\boldsymbol{\nu}}^{(k)}$ using the following*

$$\widehat{\boldsymbol{\nu}}^{(k+1)} = \arg\max_{\boldsymbol{\nu}} \sum_{i=1}^{n} \log\left(\sum_{j=1}^{m} \pi_j^{(k+1)} g(y_i; \boldsymbol{x}_i, \boldsymbol{\beta}_j^{(k+1)}, \widehat{\sigma}_j^{2(k+1)}, \widehat{\lambda}_j^{(k+1)}, \boldsymbol{\nu})\right)$$

where $g(\cdot \mid x, \boldsymbol{\theta}_j)$ is defined in (2-12).

In Section 2. 2. 5 and Section 2. 2. 6, we give a review of robust mixture regression modeling by cluster weighted modeling.

Definition 2. 3: Let $(\boldsymbol{X}, Y)$ be a pair of random variables, namely a vector of covariates $\boldsymbol{X}$ and a response variable $\boldsymbol{Y}$ defined on Ω and $\{(\boldsymbol{x}_i, y_i)\}_{i=1}^{n}$ represents a i. i. d. random sample of size n, drawn from $(\boldsymbol{X}, Y)$. Let $p(\boldsymbol{x}, y)$ be partitioned into m groups, say Ω_1, …, Ω_m. The cluster weighted models (CWM) are mixture models having density of the form

$$p(\boldsymbol{x}, y; \boldsymbol{\theta}) = \sum_{j=1}^{m} p(y|\boldsymbol{x}; \xi_j) p(\boldsymbol{x}; \psi_j) \pi_j$$

where $p(y \mid \boldsymbol{x}; \xi_j)$ is the conditional density of Y given $\boldsymbol{x}$ in Ω_j, $p(\boldsymbol{x}; \psi_j)$ is the marginal density of $\boldsymbol{X}$ in Ω_j and π_j is the weight of Ω_j in the mixture (with $\pi_j >$ 0 and $\sum_{j=1}^{m} \pi_j = 1$).

Note that the covariates here are considered random and are jointly modelled with the response variable. By also modeling the random covariates, we are able to detect outliers in the $\boldsymbol{x}$ direction and thus deal with high leverage outliers. In addition, the model estimate can also provide local distributions for the covariates.

2.2.5 Robust mixture regression with random covariates, via trimming and constraints

García-Escudero et al. (2017) proposed a new robust estimator for mixtures of regression based on the trimmed cluster weighted restricted model (CWRM). The conditional relationship between Y and $\mathbf{x}$ in the jth group is assumed as $Y = \mathrm{b}'_g\mathbf{x} + b_g^0 + \epsilon_j$ where $\epsilon_j \sim N(0, \sigma_j^2)$. The linear gaussian CWM has density of the form

$$p(\boldsymbol{x}, y; \boldsymbol{\theta}) = \sum_{j=1}^{m} \phi\left(y; \mathbf{b}_j^{'}\boldsymbol{x} + b_j^0, \sigma_j^2\right) \phi_d(\boldsymbol{x}; \boldsymbol{\mu}_j, \Sigma_j)\pi_j \tag{2-13}$$

where $\phi_d(\cdot; \boldsymbol{\mu}_j, \Sigma_j)$ denotes the density of the d-variate gaussian distribution with mean vector μ_j and covariance matrix Σ_j.

The trimmed CWRM methodology is based on the use of the log-likelihood target function to be maximized as

$$\sum_{i=1}^{n} z(\boldsymbol{x}_i, y_i) \times \log\left[\sum_{j=1}^{m} \phi\left(y_i; \mathbf{b}_j^{'}\boldsymbol{x}_i + b_j^0, \sigma_j^2\right) \phi_d(\boldsymbol{x}_i; \boldsymbol{\mu}_j, \Sigma_j)\pi_j\right] \tag{2-14}$$

where $z(\cdot, \cdot)$ is 0-1 trimming indicator function that tell us whether observation $(\boldsymbol{x}_i, y_i)$ is trimmed off $(z(\boldsymbol{x}_i, y_i) = 0)$ or not $(z(x_i, y_i) = 1)$. A fixed fraction α of observations can be unassigned be setting $\sum_{i=1}^{n} z(\mathbf{x}_i, y_i) = [n(1 -$

α)] where[·]is denoted by the largest integer that is less than or equal to the argument.

Moreover, two further constraints on the maximization in (2-14) are introduced. The first one concerns the set of eigenvalues $\{\lambda_l(\Sigma_j)\}_{l=1,\dots,d}$ of the scatter matrices Σ_j by imposing

$$\lambda_{l_1}(\Sigma_{j_1}) \leqslant c_X \lambda_{l_2}(\Sigma_{j_2}) \text{ for every } 1 \leqslant l_1 \neq l_2 \leqslant d \text{ and } 1 \leqslant j_1 \neq j_2 \leqslant m \tag{2-15}$$

The second constraint refers to the variances σ_j^2 of the regression error terms, by requiring

$$\sigma_{j_1}^2 \leqslant c_\epsilon \sigma_{j_2}^2 \text{ for every } 1 \leqslant j_1 \neq j_2 \leqslant m \tag{2-16}$$

The constraints c_X and c_ϵ, in (2-15) and (2-16), respectively, are finite real numbers, such that $c_X \geqslant 1$, $c_\epsilon \geqslant 1$. They automatically guarantee that the $|\Sigma_j| \to 0$ and $\sigma_j^2 \to 0$ cases are avoided.

The following trimmed EM algorithm was proposed by García-Escudero et al. (2017) to maximize (2-14).

Algorithm 2.6: *Given the initial parameter estimate*

$\boldsymbol{\theta}^{(0)} = (\pi_1^{(0)}, \cdots, \pi_m^{(0)}, \boldsymbol{\mu}_1^{(0)}, \dots, \boldsymbol{\mu}_m^{(0)}, \Sigma_1^{(0)}, \dots, \Sigma_m^{(0)}, b_1^{0(0)}, \dots, b_m^{0(0)}, \boldsymbol{b}_1^{(0)}, \dots, \boldsymbol{b}_m^{(0)}, \sigma_1^{2(0)}, \dots, \sigma_m^{2(0)})$, *we iterate the following E-step, C-step and M-step.*

E-step and C-steps: *Let $\boldsymbol{\theta}^{(k)}$ be the parameters at iteration k, Compute $D_i = D(\boldsymbol{x}_i, y_i; \boldsymbol{\theta}^{(k)})$ for $i=1, \dots, n$, where $D(\boldsymbol{x}_i, y_i; \boldsymbol{\theta}) = \sum_{j=1}^m \phi(y_i; \boldsymbol{b}'_j \boldsymbol{x}_i + b_j^0, \sigma_j^2)\phi_d(\boldsymbol{x}_i; \boldsymbol{\mu}_j, \Sigma_j)\pi_j$. After sorting these values, the notation $D_{(1)} \leqslant \dots \leqslant D_{(n)}$ is adopted. Consider the subset of indices $I \subset \{1, 2, \dots, n\}$ defined as $I = \{i: D_{(i)} \geqslant$*

$D_{[n\alpha]}\}$. *To update the parameters, we will take into account only the observations with indices in* I, *by setting* $\tau_{ij}^{(k)} = D_j(\boldsymbol{x}_i, y_i; \boldsymbol{\theta}^{(k)})/D(\boldsymbol{x}_i, y_i; \boldsymbol{\theta}^{(k)})$ *for* $i \in I$ *and* $\tau_{ij}^{(k)} = 0$ *otherwise.*

M-step: *Update the parameters from* τ_{ij} *using the following formulas*

$$\pi_j^{(k+1)} = \frac{\sum_{i=1}^n \tau_{ij}^{(k)}}{[n(1-\alpha)]}$$

$$\boldsymbol{\mu}_j^{(k+1)} = \frac{\sum_{i=1}^n \tau_{ij}^{(k)} \boldsymbol{x}_i}{\sum_{i=1}^n \tau_{ij}^{(k)}}$$

$$T_j = \frac{\sum_{i=1}^n \tau_{ij}^{(k)} (\boldsymbol{x}_i - \boldsymbol{\mu}_j^{(k+1)})(\boldsymbol{x}_i - \boldsymbol{\mu}_j^{(k+1)})'}{\sum_{i=1}^n \tau_{ij}^{(k)}}$$

$$b_j^{(k+1)} = \frac{\left(\frac{\sum_{i=1}^n \tau_{ij}^{(k)} y_i \boldsymbol{x}_i'}{\sum_{i=1}^n \tau_{ij}^{(k)}} - \frac{\sum_{i=1}^n \tau_{ij}^{(k)} y_i}{\sum_{i=1}^n \tau_{ij}^{(k)}} \cdot \frac{\sum_{i=1}^n \tau_{ij}^{(k)} \boldsymbol{x}_i'}{\sum_{i=1}^n \tau_{ij}^{(k)}}\right)}{\left(\frac{\sum_{i=1}^n \tau_{ij}^{(k)} \boldsymbol{x}_i \boldsymbol{x}_i'}{\sum_{i=1}^n \tau_{ij}^{(k)}} - \left(\frac{\sum_{i=1}^n \tau_{ij}^{(k)} \boldsymbol{x}_i'}{\sum_{i=1}^n \tau_{ij}^{(k)}}\right)^2\right)}$$

$$b_j^{0(k+1)} = \frac{\sum_{i=1}^n \tau_{ij}^{(k)} y_i}{\sum_{i=1}^n \tau_{ij}^{(k)}} - \left(b_j^{(k+1)}\right)' \frac{\sum_{i=1}^n \tau_{ij}^{(k)} \boldsymbol{x}_i'}{\sum_{i=1}^n \tau_{ij}^{(k)}}$$

$$s_j^2 = \frac{\sum_{i=1}^n \tau_{ij}^{(k)} (y_i - (b_j^{(k+1)})' \boldsymbol{x}_i - b_j^{0(k+1)})^2}{\sum_{i=1}^n \tau_{ij}^{(k)}}$$

where $j = 1, \ldots, m$.

To perform the constrained maximization of the sample covariance matrices, the singular-value decomposition of $T_j = U' E_j U_j$ is considered, with U_j being an orthogonal matrix and $E_j = diag(e_{j1}, e_{j2}, \ldots, e_{jd})$ a diagonal matrix. After defining the truncated eigenvalues as $[e_{jk}]_g^X = \min(c_X \cdot g, max(e_{jk}, g))$ with g being some threshold value, then the scatter matrices are finally updated as $\Sigma_j^{(k+1)} =$

$U'_j E_j^* U_j$, with

$$E_j^* = diag\left([e_{j1}]_{g_{opt}^X}^X, [e_{j2}]_{g_{opt}^X}^X, \ldots, [e_{jp}]_{g_{opt}^X}^X\right)$$

and g_{opt}^X minimizing the real valued function

$$g \longmapsto \sum_{j=1}^{m} \pi_j^{(k+1)} \sum_{k=1}^{d} \left(\log\left([e_{jk}]_g^X\right) + \frac{e_{jk}}{[e_{jk}]_g^X} \right) \tag{2-17}$$

Similarly, in case that the s_j^2 parameters do not satisfy the constraint (2-16), we consider the truncated variances $[s_j^2]_g^\epsilon = \min(c_\epsilon \cdot g, max(s_j^2, g))$. The variances of the error terms are finally updated as $\sigma_j^{2(k+1)} = [s_j^2]_{g_{opt}^\epsilon}^\epsilon$, with g_{opt}^ϵ minimizing the real valued function

$$g \longmapsto \sum_{j=1}^{m} \pi_j^{(k+1)} \left(\log\left([s_j^2]_g^\epsilon\right) + \frac{s_j^2}{[s_j^2]_g^\epsilon} \right) \tag{2-18}$$

Analogously as before, g_{opt}^ϵ can be obtained by evaluating $2m+1$ times the function(2-18).

García-Escudero et al. (2017) proposed an adaptive way of trimming to gain protection against bad leverage points. However, the involved parameters in this approach are interrelated. For instance, a high trimming level α could lead to smaller m values, since components with fewer observations may be trimmed off. Moreover, larger values of $c_{\mathbf{X}}$ and c_ϵ could lead to higher values of m since more components with few observations, but close to collinearity, may be detected.

2.2.6 Robust clustering in regression analysis via the contaminated gaussian cluster weighted model

Punzo and McNicholas (2017) proposed robust estimation of mixture regression via the contaminated gaussian cluster weighted model (CWM). Given a real-valued random vector $\mathbf{W}$, which is composed of a $d_{\mathbf{Y}}$-variate response vector $\mathbf{Y}$ and a random vector of covariates $\mathbf{X}$ of dimension $d_{\mathbf{X}}$, a contaminated gaussian distribution is given by

$$f(\mathbf{w};\boldsymbol{\mu}_{\mathbf{W}},\boldsymbol{\Sigma}_{\mathbf{W}},\alpha_{\mathbf{W}},\eta_{\mathbf{W}})=\alpha_{\mathbf{W}}\phi(\mathbf{w};\boldsymbol{\mu}_{\mathbf{W}},\boldsymbol{\Sigma}_{\mathbf{W}})+(1-\alpha_{\mathbf{W}})\phi(\mathbf{w};\boldsymbol{\mu}_{\mathbf{W}},\boldsymbol{\Sigma}_{\mathbf{W}}) \tag{2-19}$$

where $\alpha_{\mathbf{W}} \in (0, 1)$ and $\eta_{\mathbf{W}} > 1$. In (2-19), $\eta_{\mathbf{W}}$ denotes the degree of contamination.

Based on the definition of contaminated gaussian distribution, the conditional distribution of $Y|x$ from a contaminated gaussian CWM is:

$$p(\mathbf{y}|\mathbf{x};\boldsymbol{\vartheta})=\frac{\sum_{j=1}^{m}\pi_j f(\boldsymbol{x};\boldsymbol{\mu}_{\boldsymbol{x}|j},\boldsymbol{\Sigma}_{\boldsymbol{x}|j},\alpha_{\boldsymbol{x}|j},\eta_{\boldsymbol{x}|j})f(\mathbf{y};\boldsymbol{\mu}_{\boldsymbol{y}}(\boldsymbol{x};\boldsymbol{\beta}_j),\boldsymbol{\Sigma}_{\boldsymbol{y}|j},\alpha_{\boldsymbol{y}|j},\eta_{\boldsymbol{y}|j})}{\sum_{h=1}^{m}\pi_h f(\boldsymbol{x};\boldsymbol{\mu}_{\boldsymbol{x}|h},\boldsymbol{\Sigma}_{\boldsymbol{x}|h},\alpha_{\boldsymbol{x}|h},\eta_{\boldsymbol{x}|h})} \tag{2-20}$$

In this case, there are three sources of incompleteness. The first source is governed by an indicator $\boldsymbol{z}_i=(z_{i1}, \ldots, z_{im})$, where $z_{ij}=1$ if $(\boldsymbol{x}_i, \boldsymbol{y}_i)$ comes from component j and $z_{ij}=0$ otherwise. For the other two sources, use $\boldsymbol{u}_i=(u_{i1}, \ldots, u_{im})$, where $u_{ij}=1$ if $(\boldsymbol{x}_i, \boldsymbol{y}_i)$ is not an outlier in component j and $u_{ij}=0$ otherwise, and $\boldsymbol{v}_i=(v_{i1}, \ldots, v_{im})$, where $v_{ij}=1$ if $(\boldsymbol{x}_i, \boldsymbol{y}_i)$ is not a leverage point in component j and $v_{ij}=0$ otherwise. The complete-data likelihood can be written as

$$l_c(\boldsymbol{\vartheta})=l_{1c}(\boldsymbol{\pi})+l_{2c}(\boldsymbol{\alpha}_{\boldsymbol{x}})+l_{3c}(\boldsymbol{\mu}_{\boldsymbol{x}},\boldsymbol{\Sigma}_{\boldsymbol{x}},\boldsymbol{\eta}_{\mathbf{X}})+l_{4c}(\boldsymbol{\alpha}_{\boldsymbol{y}})+l_{5c}(\boldsymbol{\beta},\boldsymbol{\Sigma}\ \ ,\boldsymbol{\eta}_{\boldsymbol{y}}) \tag{2-21}$$

where $\boldsymbol{\pi}=(\pi_1, \ldots, \pi_m)$, $\boldsymbol{\mu}_X=(\boldsymbol{\mu}_{X|1}, \ldots, \boldsymbol{\mu}_{X|m})$, $\Sigma_X=(\Sigma_{X|1}, \ldots, \Sigma_{X|m})$, $\boldsymbol{\alpha}_X=(\boldsymbol{\alpha}_{X|1}, \ldots, \boldsymbol{\alpha}_{X|m})$, $\boldsymbol{\eta}_X=(\boldsymbol{\eta}\mathbf{X}_{X|1}, \ldots, \boldsymbol{\eta}_{\mathbf{X}|m})$, $\boldsymbol{\beta}=(\boldsymbol{\beta}_1, \ldots, \boldsymbol{\beta}_m)$, $\Sigma_Y=(\Sigma_{Y|1}, \ldots, \Sigma_{X|m})$, $\boldsymbol{\alpha}_Y=(\boldsymbol{\alpha}_{Y|1}, \ldots, \boldsymbol{\alpha}_{Y|m})$, $\boldsymbol{\eta}_Y=(\boldsymbol{\eta}_{Y|1}, \ldots,$

$\boldsymbol{\eta}_{Y|m}$),

$$l_{1c}(\boldsymbol{\pi}) = \sum_{i=1}^{n}\sum_{j=1}^{m} z_{ij} ln(\pi_j)$$

$$l_{2c}(\boldsymbol{\alpha_x}) = \sum_{i=1}^{n}\sum_{j=1}^{m} z_{ij}\left[v_{ij} ln(\boldsymbol{\alpha}_{\boldsymbol{x}|j}) + (1 - v_{ij}) ln(1 - \boldsymbol{\alpha}_{\boldsymbol{x}|j})\right]$$

$$l_{3c}(\boldsymbol{\mu_x}, \boldsymbol{\Sigma_x}, \boldsymbol{\eta_x}) = -\frac{1}{2}\sum_{i=1}^{n}\sum_{j=1}^{m}\{z_{ij} ln|\boldsymbol{\Sigma}_{\boldsymbol{x}|j}| + d_{\boldsymbol{x}} z_{ij}(1 - v_{ij}) ln(\boldsymbol{\eta}_{\boldsymbol{x}|j})\} - \frac{1}{2}\sum_{i=1}^{n}\sum_{j=1}^{m}\left\{z_{ij}\left(v_{ij} + \frac{1 - v_{ij}}{\boldsymbol{\eta}_{\boldsymbol{x}|j}}\right)\delta(\boldsymbol{x}_i, \boldsymbol{\mu}_{\boldsymbol{x}|j}; \boldsymbol{\Sigma}_{\boldsymbol{x}|j})\right\}$$

$$l_{4c}(\boldsymbol{\alpha_y}) = \sum_{i=1}^{n}\sum_{j=1}^{m} z_{ij}[u_{ij} ln(\boldsymbol{\alpha}_{\boldsymbol{y}|j}) + (1 - u_{ij}) ln(1 - \boldsymbol{\alpha}_{\boldsymbol{y}|j})]$$

$$l_{5c}(\boldsymbol{\beta}, \boldsymbol{\Sigma_y}, \boldsymbol{\eta_y}) = -\frac{1}{2}\sum_{i=1}^{n}\sum_{j=1}^{m}\{z_{ij} ln|\boldsymbol{\Sigma}_{\boldsymbol{y}|j}| + d_{\boldsymbol{y}} z_{ij}(1 - u_{ij}) ln(\boldsymbol{\eta}_{\boldsymbol{y}|j})\} - \frac{1}{2}\sum_{i=1}^{n}\sum_{j=1}^{m}\left\{z_{ij}\left(u_{ij} + \frac{1 - u_{ij}}{\boldsymbol{\eta}_{\boldsymbol{y}|j}}\right)\delta(\boldsymbol{y}_i, \boldsymbol{\mu_y}(\boldsymbol{x}_i; \boldsymbol{\beta}_j); \boldsymbol{\Sigma}_{\boldsymbol{y}|j})\right\}$$

and where $\delta(w, \boldsymbol{\mu}; \Sigma) = (w - \boldsymbol{\mu})'\Sigma^{-1}(w - \boldsymbol{\mu})$ denotes the squared Mahalanobis distance between w and $\boldsymbol{\mu}$, with covariance matrix Σ.

The proposed ECM algorithm iterates between three steps: an E-step and two CM-steps, until convergence. The usual parameter $\boldsymbol{\vartheta}$ is partitioned into $\boldsymbol{\vartheta}_1$ and $\boldsymbol{\vartheta}_2$, where $\boldsymbol{\vartheta}_1 = (\boldsymbol{\pi}, \boldsymbol{\mu}_X, \Sigma_X, \alpha_X, \boldsymbol{\beta}, \Sigma_Y, \boldsymbol{\alpha}_Y)$ and $\boldsymbol{\vartheta}_2 = (\boldsymbol{\eta}_X, \boldsymbol{\eta}_Y)$.

Algorithm 2. 7: ***E-step***: *Calculate* $E_{\boldsymbol{\vartheta}^{(k)}}(Z_{ij}|\boldsymbol{x}_i, \boldsymbol{y}_i)$, $E_{\boldsymbol{\vartheta}^{(k)}}(V_{ij}|\boldsymbol{x}_i, z_i)$, *and* $E_{\boldsymbol{\vartheta}^{(k)}}(U_{ij}|\boldsymbol{y}_i, z_i)$, *$i=1, \ldots, n$ and $j=1, \ldots, m$ on the $(r+1)$th iteration of the ECM algorithm, using the following formulas*:

$$E_{\boldsymbol{\vartheta}^{(k)}}(Z_{ij}|\boldsymbol{x}_i, \boldsymbol{y}_i) = \frac{\pi_j^{(k)} f\left(\boldsymbol{y}_i; \boldsymbol{\mu}_{\boldsymbol{Y}}\left(\boldsymbol{x}_i; \boldsymbol{\beta}_j^{(k)}\right), \boldsymbol{\Sigma}_{\mathbf{Y}|j}^{(k)}, \alpha_{\mathbf{Y}|j}^{(k)}, \eta_{\mathbf{Y}|j}^{(k)}\right) f\left(\boldsymbol{x}_i; \boldsymbol{\mu}_{\mathbf{X}|j}^{(k)}, \boldsymbol{\Sigma}_{\mathbf{X}|j}^{(k)}, \alpha_{\mathbf{X}|j}^{(k)}, \eta_{\mathbf{X}|j}^{(k)}\right)}{p\left(\boldsymbol{x}_i, \boldsymbol{y}_i; \boldsymbol{\vartheta}^{(k)}\right)}$$

$$=: z_{ij}^{(k)}$$

$$E_{\boldsymbol{\vartheta}^{(k)}}(V_{ij}|\boldsymbol{x}_i, \boldsymbol{z}_i) = \frac{\alpha_{\mathbf{X}|j}^{(k)} \phi\left(\boldsymbol{x}_i; \boldsymbol{\mu}_{\mathbf{X}|j}^{(k)}, \boldsymbol{\Sigma}_{\mathbf{X}|j}^{(k)}\right)}{f\left(\boldsymbol{x}_i; \boldsymbol{\mu}_{\mathbf{X}|j}^{(k)}, \boldsymbol{\Sigma}_{\mathbf{X}|j}^{(k)}, \alpha_{\mathbf{X}|j}^{(k)}, \eta_{\mathbf{X}|j}^{(k)}\right)} =: v_{ij}^{(k)}$$

and

$$E_{\boldsymbol{\vartheta}^{(k)}}(U_{ij}|\boldsymbol{y}_i, \boldsymbol{z}_i) = \frac{\alpha_{\mathbf{Y}|j}^{(k)} \phi\left(\boldsymbol{y}_i; \boldsymbol{\mu}_{\boldsymbol{Y}}\left(\boldsymbol{x}_i; \boldsymbol{\beta}_j^{(k)}\right), \boldsymbol{\Sigma}_{\mathbf{Y}|j}^{(k)}\right)}{f\left(\boldsymbol{y}_i; \boldsymbol{\mu}_{\boldsymbol{Y}}\left(\boldsymbol{x}_i; \boldsymbol{\beta}_j^{(k)}\right), \boldsymbol{\Sigma}_{\mathbf{Y}|j}^{(k)}, \alpha_{\mathbf{Y}|j}^{(k)}, \eta_{\mathbf{Y}|j}^{(k)}\right)} =: u_{ij}^{(k)}$$

***CM-step 1**: update $\hat{\boldsymbol{\vartheta}}_1^{(k+1)}$ using the following closed form expressions*:

$$\pi_j^{(k+1)} = n_j^{(k)}/n, \; \alpha_{\mathbf{X}|j}^{(k+1)} = \frac{1}{n_j^{(k)}} \sum_{i=1}^n z_{ij}^{(k)} v_{ij}^{(k)}$$

$$\boldsymbol{\mu}_{\mathbf{X}|j}^{(k+1)} = \frac{1}{s_j^{(k)}} \sum_{i=1}^n z_{ij}^{(k)} \left(v_{ij}^{(k)} + \frac{1 - v_{ij}^{(k)}}{\eta_{\mathbf{X}|j}^{(k)}}\right) \boldsymbol{x}_i \tag{2-22}$$

$$\boldsymbol{\Sigma}_{\mathbf{X}|j}^{(k+1)} = \frac{1}{n_j^{(k)}} \sum_{i=1}^n z_{ij}^{(k)} \left(v_{ij}^{(k)} + \frac{1 - v_{ij}^{(k)}}{\eta_{\mathbf{X}|j}^{(k)}}\right) \left(\boldsymbol{x}_i - \boldsymbol{\mu}_{\mathbf{X}|j}^{(k+1)}\right) \left(\boldsymbol{x}_i - \boldsymbol{\mu}_{\mathbf{X}|j}^{(k+1)}\right)' \tag{2-23}$$

$$\alpha_{\mathbf{Y}|j}^{(k+1)} = \frac{1}{n_j^{(k)}} \sum_{i=1}^n z_{ij}^{(k)} u_{ij}^{(k)} \tag{2-24}$$

$$\boldsymbol{\beta}_j^{(k+1)} = \left[\sum_{i=1}^n z_{ij}^{(k)} \left(u_{ij}^{(k)} + \frac{1 - u_{ij}^{(k)}}{\eta_{\mathbf{Y}|j}^{(k)}}\right) \boldsymbol{x}_i^* \boldsymbol{x}_i^{*'}\right]^{-1} \left[\sum_{i=1}^n z_{ij}^{(k)} \left(u_{ij}^{(k)} + \frac{1 - u_{ij}^{(k)}}{\eta_{\mathbf{Y}|j}^{(k)}}\right) \boldsymbol{x}_i^* \boldsymbol{y}_i^{*'}\right] \tag{2-25}$$

$$\boldsymbol{\Sigma}_{\mathbf{Y}|j}^{(k+1)}=\frac{1}{n_j^{(k)}}\sum_{i=1}^{n}z_{ij}^{(k)}\left(u_{ij}^{(k)}+\frac{1-u_{ij}^{(k)}}{\eta_{\mathbf{Y}|j}^{(k)}}\right)\left[\boldsymbol{y}_i-\boldsymbol{\mu}_{\boldsymbol{Y}}\left(\boldsymbol{x}_i;\boldsymbol{\beta}_j^{(k+1)}\right)\right]\left[\boldsymbol{y}_i-\boldsymbol{\mu}_{\boldsymbol{Y}}\left(\boldsymbol{x}_i;\boldsymbol{\beta}_j^{(k+1)}\right)\right]' \tag{2-26}$$

CM-step 2: *update* $\boldsymbol{\vartheta}_2^{(k+1)}$ *by maximizing*

$$-\frac{d_{\boldsymbol{X}}}{2}\sum_{i=1}^{n}z_{ij}^{(k)}\left(1-v_{ij}^{(k)}\right)ln(\eta_{\boldsymbol{X}|j})-\frac{1}{2}\sum_{i=1}^{n}z_{ij}^{(k)}\frac{1-v_{ij}^{(k)}}{\eta_{\boldsymbol{X}|j}}\delta\left(\boldsymbol{x}_i,\boldsymbol{\mu}_{\boldsymbol{X}|j}^{(k+1)};\boldsymbol{\Sigma}_{\boldsymbol{X}|j}^{(k+1)}\right)$$

with respect to $\eta_{\boldsymbol{X}|j}$, *under the constraint* $\eta_{\boldsymbol{X}|j}>1$, *and*

$$-\frac{d_{\boldsymbol{Y}}}{2}\sum_{i=1}^{n}z_{ij}^{(k)}\left(1-v_{ij}^{(k)}\right)ln(\eta_{\mathbf{Y}|j})-\frac{1}{2}\sum_{i=1}^{n}z_{ij}^{(k)}\frac{1-v_{ij}^{(k)}}{\eta_{\mathbf{Y}|j}}\delta\left(\boldsymbol{x}_i,\boldsymbol{\mu}_{\boldsymbol{Y}}\left(\boldsymbol{x}_i;\boldsymbol{\beta}_j^{(k+1)}\right);\boldsymbol{\Sigma}_{\mathbf{Y}|j}^{(k+1)}\right),$$

with respect to $\eta_{\boldsymbol{Y}|j}$, *under the constraint* $\eta_{\boldsymbol{Y}|j}>1$ *for each* $j=1, \dots, m$.

Based on (2-22), $\boldsymbol{\mu}_{\mathbf{X}|j}^{(k+1)}$ is a weighted mean of the $\boldsymbol{x}_i$ values, with weights depending on

$$v_{ij}^{(k)}+\frac{1-v_{ij}^{(k)}}{\eta_{\mathbf{X}|j}^{(k)}} \tag{2-27}$$

Analogously, based on (2-25), the regression coefficients $\boldsymbol{\beta}_j^{(k+1)}$ can be considered a weighted least squares estimate with weights depending on

$$u_{ij}^{(k)}+\frac{1-u_{ij}^{(k)}}{\eta_{\mathbf{Y}|j}^{(k)}} \tag{2-28}$$

The weights in (2-27) and (2-28) reduce, respectively, the effect of leverage points in the estimation of $\boldsymbol{\mu}_{\mathbf{X}|j}$ and the effect of outliers in the estimation of $\boldsymbol{\beta}_j$, thereby providing a robust way to estimate $\boldsymbol{\mu}_{X|j}$ *and* β_j, $j=1, \dots, m$. In addition, from (2-23) and (2-26), the larger squared standardized residuals also have smaller effect on $\Sigma_{\mathbf{X}|j}$ and $\Sigma_{\mathbf{X}|j}$, $j=1, \dots, m$ due to the weights in

(2-27) and (2-28).

When the contaminated gaussian CWM is used for detection of atypical points in each group, $1-\alpha_{\mathbf{X}|j}$ *and* $1-\alpha_{\mathbf{Y}|j}$ represent the proportion of leverage points and outliers, respectively. Similar to TLE, one could require that in the jth group, $j=1, \ldots, m$ the proportion of typical observations, with respect to **X** and **Y**, respectively, to be at least equal to a predetermined value α^*. However, pre-specifying points as outliers and/or leverage a priori may not be realistic in many practical scenarios.

In Section 2.2.7 and Section 2.2.8, we introduce two robust methods of trimmed estimator: Trimmed likelihood estimator (TLE) and Least trimmed squares estimator (LTS).

2.2.7 Trimmed likelihood estimator

Neykov et al. (2007) proposed to fit the mixture model using the trimmed likelihood method (TLE). The TLE is defined as

$$\widehat{\boldsymbol{\theta}}_{TLE} = \arg\max_{\boldsymbol{\theta}, S_h} \sum_{i \in S_h} \log\left\{\sum_{j=1}^{m} \pi_j \phi(y_i; \boldsymbol{x}_i^{\mathrm{T}} \boldsymbol{\beta}_j, \sigma_j^2)\right\}$$

where S_h is a subset of $(1, 2, \cdots, n)$ with the cardinality h, h is the round number of $n(1-\alpha)$ and α is percentage of the data to be trimmed.

The FAST-TLE algorithm is used by Neykov et al. (2007) to compute the $\hat{\boldsymbol{\theta}}_{TLE}$ and consists of a two-step procedure: a trial step followed by a refinement step. In the trial step several subsamples of size h^* is selected randomly from the data sample and then the model is fitted to that subsample to get multiple trial ML estimates. The trial subsample size h^* should be greater than or equal to $m(p+$

1). The refinement step is based on the following procedure: Ⅰ The cases with the h smallest negative log-likelihoods based on the current estimate are found, starting with each of the trial MLE as the initial estimator; Ⅱ Fitting the mixture regression model (2-1) to these h cases gives an improved fit. Repeating Ⅰ and Ⅱ yields an iterative process. The converged estimate with the smallest negative log-likelihood is selected as the final estimate among multiple initial estimators.

The choice of α the proportion of trimming, plays an important role for the TLE and should be predetermined. Unlike the adaptively trimmed version proposed by Yao et al. (2014), α is a fixed value. If α is too large, the TLE will lose much efficiency. If α is too small and the percentage of outliers is more than α, then the TLE will fail.

In order to improve the performance of TLE, García-Escudero et al. (2010) proposed an adaptive way of trimming and protection against bad leverage points. García-Escudero et al. (2010) stated that the first trimming α_1 in TLE is useful in avoiding the effect of outliers in y but it has almost null effect in diminishing the effect of high leverage outliers, therefore, a further "second" trimming of size α_2 was proposed taking into account only the values of the explanatory variables of the observations surviving to the first trimming. With respect to choosing α_1 and α_2, a preventive fashion is recommended, i.e., choose values for α_1 and α_2 a bit larger than needed though which all the outlying observations together with some non-outlying ones are surely removed. Afterwards, standard regression analysis tools can be applied to recover the observations that should not have been trimmed off.

2.2.8 Least trimmed squares estimator

Rousseeuw (1984) proposed the Least trimmed squares (LTS) estimator for the linear regression parameters, which is obtained as the solution of the following minimization problem:

$$\min_{\hat{\beta}} \sum_{i=1}^{h} (r^2)_{i:n} \tag{2-29}$$

where $r_i = y_i - \boldsymbol{x}_i^{\mathrm{T}}\beta$, $(r^2)_{1:n} \leqslant \cdots \leqslant (r^2)_{n:n}$ are the ordered squared residuals, $h=[n(1-\alpha)+1]$ is the number of observations after trimming, and α is the proportion of trimming.

Motivated by LTS method in linear regression, Doğru and Arslan (2017) proposed the Least trimmed squares estimator for mixture regression parameters. Note that the second term of the complete data log-likelihood function given in formula (2-4) is basically the least-squares (LS) criterion which is known to be sensitive to the outliers and should be robustified. Doğru and Arslan (2017) proposed the following adapted complete data log-likelihood function using the LTS criterion given in (2-29):

$$l_n^c(\boldsymbol{\theta}) = \sum_{i=1}^{n}\sum_{j=1}^{m} z_{ij}\left\{\log(\pi_j) - \frac{1}{2}\log(2\pi) - \frac{1}{2}\log(\sigma_j^2)\right\} - \sum_{j=1}^{m}\sum_{i \in S_{jh}} z_{ij}\frac{r_{ij}^2}{2\sigma_j^2}$$

where $r_{ij}=y_i-\boldsymbol{x}_i^{\mathrm{T}}\beta_j$, and S_{jh} is a subset of $(1, 2, \cdots, n)$ with the cardinality h such that $\{r_{ij}^2, i \in S_{jh}\}$ are the h smallest squared residuals among $\{r_{ij}^2, i=1, \ldots, n\}$ for $j=1, \ldots, m$. Note that the h observations depend on jth component in the mixture regression, i.e., different components j have different h observations.

To run the EM algorithm, the conditional expectation of the complete data log-likelihood function is taken to get rid of the latency of z_{ij}

$$E(l_n^c(\boldsymbol{\theta})|y_i) = \sum_{i=1}^{n}\sum_{j=1}^{m} E(z_{ij}|y_i)\left\{\log(\pi_j) - \frac{1}{2}\log(2\pi) - \frac{1}{2}\log(\sigma_j^2)\right\} - \sum_{j=1}^{m}\sum_{i\in S_{jh}} E(z_{ij}|y_i)\frac{r_{ij}^2}{2\sigma_j^2}$$

The conditional expectation $E(z_{ij}|y_i)$ can be calculated using the classical theory of the mixture models. Then, the steps of the EM algorithm for the mixture regression based on the LTS estimation method will be as follows.

Algorithm 2.8: *Set the initial estimator $\boldsymbol{\theta}^{(0)}$. At $(k+1)$th step the EM algorithm iterates the following E-step and M-step*:

E-step: *Calculate the classification probabilities using the same formula as in formula (2.5). Let $r_{ij}^{(k+1)} = y_i - \boldsymbol{x}_i^{\mathrm{T}}\boldsymbol{\beta}_j^{(k)}$. For each j, let $S_{jh}^{(k+1)}$ collect the indexes of observations that correspond to the h smallest values among $\{|r_{ij}^{(k+1)}|, i=1, \ldots, n\}$.*

M-step: *Update parameter estimates using the following formulas*:

$$\pi_j^{(k+1)} = (\sum_{i=1}^{n} p_{ij}^{(k+1)})/n$$

$$\boldsymbol{\beta}_j^{(k+1)} = \left(\sum_{i\in S_{jh}^{(k+1)}} \boldsymbol{x}_i\boldsymbol{x}_i^{\mathrm{T}} p_{ij}^{(k+1)}\right)^{-1}\left(\sum_{i\in S_{jh}^{(k+1)}} \boldsymbol{x}_i p_{ij}^{(k+1)} y_i\right)$$

$$\sigma_j^{2(k+1)} = c_\alpha \frac{\sum_{i\in S_{jh}^{(k+1)}} p_{ij}^{(k+1)}(y_i - \boldsymbol{x}_i^{\mathrm{T}}\boldsymbol{\beta}_j^{(k+1)})^2}{\sum_{i\in S_{jh}^{(k+1)}} p_{ij}^{(k+1)} - p}$$

where $j=1, \ldots, m$, and c_α is a consistency constant. For the normal errors, c_α will be $(1-\alpha)/F_{\chi_3^2}(q_\alpha)$ with $q_\alpha = \chi^2_{1,1-\alpha}$. Here $F_{\chi_3^2}$ is the cumulative distribution *function of the* χ^2 *distribution with the* 3 *degrees of freedom and* q_α *is the upper* α *percent point of the* χ^2 *distribution with the* 1 *degree of freedom.*

Similar to trimmed likelihood estimator (TLE), the choice of α should be predetermined and α is a fixed value. If α is too large, the LTS estimator will lose much efficiency. If α is too small and the percentage of outliers is more than α, then the LTS estimator will fail.

2.2.9 Robust estimator based on a modified EM algorithm with bisquare loss

Bai et al. (2012) developed a modified EM algorithm by replacing the least squares criterion in the M-step of Algorithm 3. 2. 1 using a robust criterion. In the M step of algorithm, $\boldsymbol{\beta}_j^{(k+1)}$, $j=1, \ldots, m$ is the solution of

$$\sum_{i=1}^{n} p_{ij}^{(k+1)} \mathbf{x}_i \psi\left(\frac{y_i - \boldsymbol{x}_i^{\mathrm{T}} \boldsymbol{\beta}_j}{\sigma^{(k)}}\right) = 0$$

where $\sigma^{(k)}$ is a robust scale estimate of the error terms and $\psi(\cdot)$ is Tukey's bisquare function $\psi_c(t) = t\{1-(t/c)^2\}_+^2$, which weights the tail contribution of t by a biweight function. In the literature of parametric robustness, the use of $c=4.685$ is recommended.

More specifically, Bai et al. (2012) proposed the following modified EM-type algorithm.

Algorithm 2. 9: *Given the initial parameter estimate* $\boldsymbol{\theta}^{(0)}$, *at* $(k+1)$ *th step, the EM algorithm iterates the following E-step and M-step*:

E-step: *Calculate the classification probabilities using the same formula as in* (2.5).

***M*–step**: *Update the parameters*

$$\boldsymbol{\beta}_j^{(k+1)} = \left(\sum_{i=1}^{n} p_{ij}^{*(k+1)} \boldsymbol{x}_i \boldsymbol{x}_i^{\mathrm{T}}\right)^{-1} \sum_{i=1}^{n} p_{ij}^{*(k+1)} \boldsymbol{x}_i y_i$$

$$\pi_j^{(k+1)} = \frac{1}{n} \sum_{i=1}^{n} p_{ij}^{(k+1)}$$

$$\sigma_j^{2(k+1)} = \frac{2 \sum_{i=1}^{n} \sum_{j=1}^{m} p_{ij}^{(k+1)} (y_i - \boldsymbol{x}_i^{\mathrm{T}} \boldsymbol{\beta}_j^{(k+1)})^2 \omega_{ij}^{(k+1)}}{\sum_{i=1}^{n} p_{ij}^{(k+1)}}$$

where $j=1, \ldots, m$, $W(t)=\psi(t)/t$,

$$p_{ij}^{*(k+1)} = p_{ij}^{(k+1)} W\left(\frac{y_i - \boldsymbol{x}_i^{\mathrm{T}} \boldsymbol{\beta}_j^{(k)}}{\sigma^{(k)}}\right)$$

and

$$\omega_{ij}^{(k+1)} = \min\left[1 - \left\{1 - \left(\frac{y_i - \boldsymbol{x}_i^{\mathrm{T}} \boldsymbol{\beta}_j^{(k+1)}}{1.56\sigma^{(k)}}\right)^2\right\}^3, 1\right] \left(\frac{\sigma^{(k)}}{y_i - \boldsymbol{x}_i^{\mathrm{T}} \boldsymbol{\beta}_j^{(k+1)}}\right)^2$$

Note, however, there is no clear objective function for the modified EM algorithm proposed by Bai et al. (2012). When there are multiple roots to the estimating equations starting from many random initial values, Bai et al. (2012) proposed to choose the modal root, i.e., the root which most of the initial values converge to. The empirical study demonstrates that the proposed modal root is effective when multiple solutions were found and can be also applied to other estimating equation problems. However, it requires more research to provide some theoretical guideline for the choice of a consistent root of an estimating equation.

2.2.10 Robust EM-type algorithm for log-concave mixtures of regression models

Hu et al. (2017) proposed a robust EM-type algorithm by assuming the component error densities are log-concave. A density $g(x)$ is log-concave if its log-density, $\phi(x)=\log g(x)$, is concave. Examples of log-concave densities include, but are not limited to normal, laplace, chi-square, logistic, gamma with shape parameter greater than 1, and beta distribution with both parameters greater than 1. For robust mixture regression models, it is assumed that error distribution g_j's are log-concave, i.e., $\log g_j$ is concave, for $j=1, \ldots, m$. The likelihood function for the mixture of regression model can be presented as:

$$f(y_i|\boldsymbol{x}_i, \boldsymbol{\Psi}, \mathrm{g}) = \sum_{j=1}^{m} \pi_j g_j(y_i - \boldsymbol{x}_i^{\mathrm{T}}\boldsymbol{\beta}_j), \qquad i=1,\ldots,n$$

where $\Psi=(\pi_1, \beta_1, \ldots, \pi_m, \beta_m)^{\mathrm{T}}$.

The unknown parameter $\boldsymbol{\Psi}$ can be estimated by maximizing the observed log-likelihood function:

$$\ell(\boldsymbol{\Psi}, \mathrm{g}|\boldsymbol{X}, \mathrm{y}) = \sum_{i=1}^{n} \log \sum_{j=1}^{m} \pi_j g_j(y_i - \boldsymbol{x}_i^{\mathrm{T}}\boldsymbol{\beta}_j) \tag{2-30}$$

where $g_j(x)=\exp\{\phi_j(x)\}$ for some unknown concave function $\phi_j(x)$.

The proposed robust EM-type algorithm by Hu et al. (2017) to maximize (2-30) is as follows.

Algorithm 2.10: *Given the initial parameter estimate* $\Psi^{(0)}$ *and* $g^{(0)}$, *iterate the following E-step and M-step.*

E-step: *Calculate the classification probabilities*:

$$p_{ij}^{(k+1)} = \frac{\pi_j^{(k)} g_j^{(k)}(y_i - \boldsymbol{x}_i^{\mathrm{T}} \boldsymbol{\beta}_j^{(k)})}{\sum_{j=1}^{m} \pi_j^{(k)} g_j^{(k)}(y_i - \boldsymbol{x}_i^{\mathrm{T}} \boldsymbol{\beta}_j^{(k)})}.$$

***M-step*:**

(*A*) *Update* π_j*'s using the formula*: $\pi_j^{(k+1)} = \frac{1}{n} \sum_{i=1}^{n} p_{ij}^{(k+1)}$, $j = 1, \ldots, m$.

(*B*) *Update* $\boldsymbol{\beta}$:

$$\tilde{\boldsymbol{\beta}}_j^{(k+1)} \leftarrow \arg\max_{\boldsymbol{\beta}_j} \sum_{i=1}^{n} p_{ij}^{(k+1)} \log g_j^{(k)}(y_i - \boldsymbol{x}_i^{\mathrm{T}} \boldsymbol{\beta}_j), \qquad j = 1, \ldots, m.$$

(*C*) *Shift the intercept of* $\tilde{\boldsymbol{\beta}}_j^{(k+1)}$ *so that the residuals have a zero mean.*

$$\hat{\boldsymbol{\beta}}_j^{(k+1)} = (\hat{\beta}_{j,0}^{(k+1)}, \tilde{\boldsymbol{\beta}}_{j,1}^{(k+1)}, \ldots, \tilde{\boldsymbol{\beta}}_{j,p-1}^{(k+1)}),$$

$$\begin{aligned} \textit{where } \hat{\beta}_{j,0}^{(k+1)} &= \tilde{\beta}_{j,0}^{(k+1)} + c_j^{(k+1)} \textit{ with } c_j^{(k+1)} \\ &= \tfrac{1}{n} \textstyle\sum_{i=1}^{n} p_{ij}^{(k+1)} (y_i - \boldsymbol{x}_i^{\mathrm{T}} \tilde{\boldsymbol{\beta}}_j), \textit{ for } j = 1, \ldots, m. \end{aligned}$$

(*D*) *Update* g_j *by*:

$$g_j^{(k+1)} \leftarrow \arg\max_{g_j \in \mathbb{G}} \sum_{i=1}^{n} p_{ij}^{(k+1)} \log g_j(y_i - \boldsymbol{x}_i^{\mathrm{T}} \hat{\boldsymbol{\beta}}_j^{(k+1)}), \qquad j = 1, \ldots, m. \tag{2-31}$$

where $\mathbb{G}$ *is the family of all log-concave densities.*

In (2-31), the error density g can be updated by the R package LogConcDEAD based on fitted residuals $y_i - \boldsymbol{x}_i^{\mathrm{T}} \boldsymbol{\beta}_j^{(k+1)}$.

Unlike the methods mentioned in the previous sections, the mixture regression model proposed by Hu et al. (2017) is semiparametric in the sense that it does not require a specific parametric assumption of error densities and can be adaptive to different error densities with weak log-concave assumption.

2.3 Simulation studies

In this section, we use a simulation study to demonstrate the effectiveness of some of the reviewed methods and compare them with MLE. Due to the limited space, we can not include all of reviewed methods for comparison in the simulation study. We mainly choose some methods that are popularly used and can be found from the existing R packages. We will consider the following six methods: Ⅰ traditional MLE assuming the error has normal density (MLE); Ⅱ the robust mixture regression based on t-distribution (Mixregt, Yao et al. 2014); Ⅲ the robust mixture regression based on Laplace distribution (MixregL, Song et al., 2014); Ⅳ trimmed likelihood estimator (TLE, Neykov et al., 2007) with the percentage of trimmed data α set to 0.1; Ⅴ the robust modified EM algorithm based on bisquare (MEM-bisquare, Bai et al. 2012); Ⅵ the trimmed cluster weighted restricted model (CWRM, García-Escudero et al. 2017).

To compare different methods, we report the mean squared errors (MSE) and absolute robust bias (RB) of the parameter estimates for each estimation method, where robust bias is defined as:

$$RB_j = \text{median}(\hat{\beta}_j) - \beta_j$$

Note, however, for mixture models, there are well known label switching issues (Celeux et al., 2000; Stephens, 2000; Yao and Lindsay, 2009; Yao, 2012) when doing comparison using the simulation study. There are no widely accepted labeling methods. In our simulation study, we choose the labels by minimizing the distance to the true parameter values. However, it requires more research to compare different labeling methods.

Example 2.1: In this example, we consider a case with two components. For each $i=1, \ldots, n$ y_i is independently generated with

$$y_i=\begin{cases}1-x_{i1}+x_{i2}+\gamma_i\sigma+\epsilon_{i1}, & if\ z_{i1}=1\\ 1+3x_{i1}+x_{i2}+\gamma_i\sigma+\epsilon_{i2}, & if\ z_{i1}=0\end{cases}$$

where $\sigma=1$, z_{i1} is a component indicator generated from a Bernoulli distribution with $P(z_{i1}=1)=0.3$, x_{i1} and x_{i2} are independently generated from $N(0, 1)$, and the error terms ϵ_{i1} and ϵ_{i2} have the same distribution as ϵ. In order to generate outliers into the data, we introduce a mean-shift parameter, γ_i, to the mixture regression model (see details in chapter 4). We consider the following six cases for the error density of ϵ:

Case Ⅰ: $\epsilon \sim N(0, 1)$ -standard normal distribution.

Case Ⅱ: $\epsilon \sim t_3$-t-distribution with degree of freedom 3.

Case Ⅲ: $\epsilon \sim t_1$ - t - distribution with degree of freedom 1 (Cauchy distribution).

Case Ⅳ: $\epsilon \sim N(0, 1)$ with 5% of low leverage outliers.

Case Ⅴ: $\epsilon \sim N(0, 1)$ with 5% of high leverage outliers.

Case Ⅵ: $\epsilon \sim N(0, 1)$ with 20% of gross outliers where outliers are concentrated in a very remote position.

In Case Ⅰ, Case Ⅱ and Case Ⅲ, mean-shift parameter γ_i's are set to be zero. In Case Ⅳ and Case Ⅴ, 5% of nonzero mean-shift parameter γ_i is randomly generated from a uniform distribution between 11 and 13 for the second component, and 95% of γ_i's are set to be zero. In Case Ⅴ, particularly, 5% outliers are also replaced by $x_{i1}=20$, $x_{i2}=20$ and $y_i=100$. In Case Ⅵ, 20% of nonzero mean-shift parameter is randomly generated from a uniform distribution between 101 and 102 for the second component, and 80% of γ_i's are set to be zero. Case Ⅰ is used to test the efficiency of different estimation methods compared to the traditional MLE when the error is exactly normally distributed and there are no outliers. Case Ⅱ is a heavy-tailed distribution. The t-distributions with degree of freedom from 3 to 5 are often used to represent the heavy-tailed distributions. Case Ⅲ is a Cauchy distribution which has extreme heavy tails. In Case Ⅳ, the 5% data from the second component are likely to be low leverage outliers. In Case Ⅴ, the 5% outliers are high leverage outliers. In Case Ⅵ, we want to provide an empirical idea of "breakdown points" for the compared procedures in very extreme cases.

Table 2.1 and Table 2.2 report MSE(RB) of the parameter estimates for each estimation method for sample size $n=200$ and $n=400$, respectively. The number of replicates is 200. Based on Table 2.1 and Table 2.2, we have the following findings:

(1) The MLE worked the best for Case Ⅰ ($\epsilon \sim N(0, 1)$), but failed to provide reasonable estimates for Case Ⅱ to Ⅵ.

(2) CWRM and TLE work well if the trimming proportion is properly chosen. For the rest of robust methods, none of them has clear advantage over the others.

(3) All methods broke down except for CWRM in Case Ⅵ where there are 20% gross outliers in the simulated data. When the proportion of outliers is set to 10% (not reported due to the limited space), all methods worked reasonably well.

Table 2.1 ***MSE* (*RB*) *of point estimates for n = 200 in Example 2.1***

TRUE	MLE	Mixregt	MixregL	TLE	MEM-bisquare	CWRM
			Case Ⅰ: $\epsilon \sim N(0, 1)$			
β_{10} : 1	0.022(0.010)	0.055(0.009)	0.041(0.004)	0.102(0.009)	0.031(0.009)	0.025(−0.007)
β_{20} : 1	0.010(0.002)	0.017(0.017)	0.012(0.004)	0.025(0.011)	0.011(0.008)	0.010(0.011)
β_{11} : −1	0.023(0.016)	0.116(0.045)	0.039(0.032)	0.323(0.123)	0.081(0.018)	0.116(0.186)
β_{21} : 3	0.008(0.009)	0.015(0.018)	0.011(0.012)	0.016(0.033)	0.009(0.009)	0.021(−0.050)
β_{12} : 1	0.027(0.025)	0.060(0.013)	0.049(0.004)	0.082(0.005)	0.038(0.025)	0.061(−0.013)
β_{22} : 1	0.009(0.008)	0.015(0.011)	0.012(0.002)	0.014(0.006)	0.009(0.007)	0.025(−0.004)
π_1 : 0.3	0.002(0.000)	0.002(0.008)	0.002(0.003)	0.003(0.003)	0.002(0.000)	0.003(0.003)
			Case Ⅱ: $\epsilon \sim t_3$			
β_{10} : 1	5.095(0.007)	0.079(0.031)	0.070(0.033)	0.109(0.028)	0.069(0.006)	0.027(−0.007)
β_{20} : 1	2.488(0.003)	0.016(0.004)	0.016(0.005)	0.016(0.002)	0.014(0.002)	0.010(0.005)
β_{11} : −1	2.216(0.012)	0.049(0.015)	0.058(0.005)	0.160(0.084)	0.120(0.002)	0.055(0.026)
β_{21} : 3	1.074(0.001)	0.018(0.009)	0.013(0.006)	0.017(0.020)	0.014(0.018)	0.021(−0.007)
β_{12} : 1	5.528(0.003)	0.069(0.007)	0.068(0.007)	0.089(0.034)	0.072(0.012)	0.048(−0.005)
β_{22} : 1	5.678(0.020)	0.017(0.003)	0.019(0.017)	0.015(0.005)	0.014(0.011)	0.090(0.018)
π_1 : 0.3	0.015(0.002)	0.003(0.005)	0.002(0.007)	0.003(0.018)	0.004(0.021)	0.002(0.010)
			Case Ⅲ: $\epsilon \sim t_1$			
β_{10} : 1	7.2e+4(0.081)	0.175(0.018)	0.115(0.057)	1.733(0.027)	0.288(0.027)	0.138(0.014)
β_{20} : 1	2.7e+5(0.026)	0.020(0.019)	0.045(0.067)	1.711(0.013)	0.038(0.028)	0.039(0.020)
β_{11} : −1	1.9e+5(1.682)	0.251(0.033)	4.504(0.486)	1.141(0.055)	0.657(0.034)	0.163(0.208)
β_{21} : 3	1.1e+5(1.241)	0.020(0.003)	0.166(0.271)	0.185(0.006)	0.078(0.031)	0.095(0.229)
β_{12} : 1	4.0e+4(0.044)	0.142(0.016)	0.080(0.035)	1.178(0.013)	0.279(0.052)	0.170(−0.017)
β_{22} : 1	2.5e+5(0.047)	0.023(0.010)	0.030(0.002)	0.220(0.049)	0.066(0.023)	0.154(0.083)
π_1 : 0.3	0.253(0.199)	0.003(0.002)	0.019(0.045)	0.014(0.011)	0.008(0.042)	0.004(0.043)
			Case Ⅳ: $\epsilon \sim N(0, 1)$ with 5% of low leverage outliers			
β_{10} : 1	2.866(0.058)	0.052(0.058)	1.401(0.940)	0.049(0.018)	0.027(0.022)	0.023(0.019)
β_{20} : 1	119.7(11.54)	0.015(0.010)	0.936(0.053)	0.011(0.007)	0.010(0.005)	0.009(0.010)
β_{11} : −1	7.891(0.890)	0.043(0.025)	2.771(2.033)	0.046(0.074)	0.025(0.015)	0.057(0.074)

Table 2.1 ***MSE (RB) of point estimates for n = 200 in Example* 2.1 (Continued table)**

TRUE	MLE	Mixregt	MixregL	TLE	MEM-bisquare	CWRM
β_{21} : 3	0.850(0.585)	0.014(0.011)	0.273(0.293)	0.011(0.020)	0.009(0.011)	0.021(0.007)
β_{12} : 1	0.510(0.006)	0.060(0.027)	2.334(1.985)	0.057(0.002)	0.034(0.007)	0.046(0.001)
β_{22} : 1	0.368(0.005)	0.017(0.006)	0.572(0.017)	0.015(0.014)	0.011(0.015)	0.097(0.055)
π_1 : 0.3	0.362(0.635)	0.003(0.015)	0.012(0.016)	0.003(0.020)	0.003(0.033)	0.003(0.013)
			Case V: $\epsilon \sim N(0, 1)$ with 5% of high leverage outliers			
β_{10} : 1	0.090(0.002)	0.064(-0.017)	0.037(0.015)	0.079(-0.012)	0.044(0.003)	0.020(-0.012)
β_{20} : 1	0.015(0.018)	0.040(0.025)	0.020(0.033)	0.037(0.021)	0.013(0.015)	0.009(-0.007)
β_{11} : -1	3.037(1.830)	0.620(0.069)	0.031(0.056)	1.421(0.130)	0.146(0.023)	0.071(0.070)
β_{21} : 3	0.181(0.416)	0.187(0.422)	0.190(0.422)	0.123(0.325)	0.087(0.046)	0.018(-0.038)
β_{12} : 1	1.037(0.550)	0.111(-0.026)	0.050(-0.023)	0.130(-0.036)	0.111(-0.009)	0.053(0.019)
β_{22} : 1	0.276(0.517)	0.293(0.522)	0.282(0.531)	0.201(0.445)	0.106(0.107)	0.110(-0.015)
π_1 : 0.3	0.008(0.023)	0.006(0.025)	0.003(0.022)	0.008(0.033)	0.003(-0.002)	0.002(0.002)
			Case VI: 20% gross outliers			
β_{10} : 1	2.125(-0.010)	117.4(0.039)	147.3(0.030)	154.2(0.008)	13.26(-0.016)	0.019(-0.005)
β_{20} : 1	1.0e+4(101.5)	305.4(0.005)	336.3(6.290)	154.5(-0.008)	259.0(0.013)	0.010(-0.003)
β_{11} : -1	6.389(2.541)	17.72(0.075)	38.65(3.278)	0.483(0.005)	14.81(2.571)	0.040(0.015)
β_{21} : 3	0.050(-0.006)	12.70(-0.070)	187.5(9.991)	0.013(-0.004)	13.57(-1.277)	0.024(-0.017)
β_{12} : 1	0.037(-0.007)	20.69(0.001)	506.2(0.009)	0.025(0.006)	0.045(0.002)	0.040(-0.010)
β_{22} : 1	0.031(0.034)	46.75(-0.019)	994.8(30.15)	0.016(-0.018)	9.045(0.002)	0.108(-0.010)
π_1 : 0.3	0.249(0.500)	0.023(0.083)	0.184(0.470)	0.012(0.076)	0.044(0.200)	0.005(0.050)

Table 2.2 ***MSE (RB) of point estimates for n = 400 in Example* 2.1**

TRUE	MLE	Mixregt	MixregL	TLE	MEM-bisquare	CWRM
			Case I: $\epsilon \sim N(0, 1)$			
β_{10} : 1	0.012(0.008)	0.026(0.001)	0.025(0.038)	0.032(0.005)	0.013(0.002)	0.014(0.017)
β_{20} : 1	0.004(0.006)	0.007(0.001)	0.006(0.023)	0.006(0.007)	0.005(0.009)	0.005(0.002)
β_{11} : -1	0.010(0.003)	0.020(0.017)	0.013(0.023)	0.040(0.095)	0.011(0.006)	0.098(0.142)
β_{21} : 3	0.004(0.004)	0.007(0.009)	0.007(0.004)	0.008(0.009)	0.005(0.003)	0.012(0.057)

Table 2.2 *MSE* (*RB*) *of point estimates for n* =400 *in Example* 2.1 (Continued table)

TRUE	MLE	Mixregt	MixregL	TLE	MEM-bisquare	CWRM
β_{12} : 1	0.015(0.010)	0.029(0.019)	0.024(0.031)	0.032(0.028)	0.016(0.011)	0.042(0.017)
β_{22} : 1	0.005(0.013)	0.007(0.018)	0.007(0.002)	0.007(0.017)	0.005(0.012)	0.009(−0.002)
π_1 : 0.3	0.001(0.001)	0.001(0.012)	0.001(0.002)	0.001(0.010)	0.001(0.001)	0.003(−0.035)
Case Ⅱ: $\epsilon \sim t_3$						
β_{10} : 1	8.523(0.009)	0.026(0.001)	0.023(0.009)	0.031(0.002)	0.024(0.003)	0.015(0.019)
β_{20} : 1	0.097(0.014)	0.008(0.008)	0.009(0.002)	0.008(0.009)	0.007(0.003)	0.005(0.002)
β_{11} : −1	0.603(0.006)	0.029(0.008)	0.021(0.008)	0.036(0.078)	0.020(0.005)	0.041(0.092)
β_{21} : 3	0.832(0.002)	0.008(0.009)	0.008(0.006)	0.008(0.022)	0.008(0.016)	0.012(−0.022)
β_{12} : 1	4.268(0.030)	0.032(0.012)	0.031(0.020)	0.034(0.003)	0.027(0.004)	0.028(−0.014)
β_{22} : 1	1.959(0.004)	0.007(0.012)	0.009(0.010)	0.008(0.003)	0.007(0.001)	0.045(0.004)
π_1 : 0.3	0.005(0.013)	0.001(0.006)	0.001(0.003)	0.001(0.004)	0.001(0.013)	0.001(0.003)
Case Ⅲ: $\epsilon \sim t_1$						
β_{10} : 1	5.1e+4(0.032)	0.040(0.006)	0.037(0.024)	0.088(0.053)	0.055(0.027)	0.073(0.015)
β_{20} : 1	8.5e+5(0.141)	0.007(0.014)	0.010(0.029)	0.017(0.010)	0.012(0.001)	0.016(0.005)
β_{11} : −1	2.7e+4(1.171)	0.079(0.056)	4.413(0.421)	0.066(0.024)	0.421(0.024)	0.127(0.225)
β_{21} : 3	7.7e+5(0.651)	0.008(0.015)	0.207(0.270)	0.020(0.038)	0.071(0.045)	0.059(0.178)
β_{12} : 1	1.7e+5(0.112)	0.068(0.020)	0.075(0.034)	0.130(0.032)	0.144(0.020)	0.087(0.007)
β_{22} : 1	6.3e+5(0.015)	0.010(0.023)	0.013(0.039)	0.021(0.006)	0.014(0.007)	0.029(0.010)
π_1 : 0.3	0.231(0.054)	0.001(0.002)	0.022(0.069)	0.002(0.015)	0.005(0.047)	0.003(0.042)
Case Ⅳ: $\epsilon \sim N(0, 1)$ with 5% of low leverage outliers						
β_{10} : 1	0.450(0.076)	0.024(0.027)	1.709(1.042)	0.023(0.019)	0.015(0.001)	0.011(0.003)
β_{20} : 1	131.5(11.59)	0.008(0.006)	0.489(0.003)	0.007(0.002)	0.005(0.004)	0.004(0.001)
β_{11} : −1	8.385(2.916)	0.016(0.027)	2.434(2.195)	0.021(0.073)	0.011(0.021)	0.032(0.048)
β_{21} : 3	0.635(0.644)	0.008(0.011)	0.374(0.301)	0.006(0.018)	0.005(0.003)	0.010(0.021)
β_{12} : 1	0.024(0.014)	0.021(0.007)	2.575(1.551)	0.021(0.023)	0.014(0.004)	0.025(0.005)
β_{22} : 1	0.190(0.001)	0.009(0.006)	0.779(0.060)	0.008(0.006)	0.006(0.007)	0.047(0.014)
π_1 : 0.3	0.397(0.638)	0.001(0.013)	0.012(0.017)	0.001(0.017)	0.001(0.028)	0.001(0.010)

Table 2.2 ***MSE (RB) of point estimates for n = 400 in Example* 2.1 (Continued table)**

TRUE	MLE	Mixregt	MixregL	TLE	MEM-bisquare	CWRM
Case V: $\epsilon \sim N(0, 1)$ with 5% of high leverage outliers						
β_{10} : 1	0.081(−0.041)	0.022(0.032)	0.015(0.006)	0.023(0.022)	0.011(0.016)	0.011(0.002)
β_{20} : 1	0.007(0.023)	0.014(0.032)	0.010(0.035)	0.011(0.012)	0.006(0.009)	0.005(0.005)
β_{11} : −1	2.016(1.730)	0.020(0.046)	0.014(0.069)	0.297(0.082)	0.073(0.012)	0.029(0.074)
β_{21} : 3	0.178(0.419)	0.195(0.432)	0.132(0.410)	0.124(0.365)	0.074(0.036)	0.008(0.029)
β_{12} : 1	0.450(0.480)	0.024(0.006)	0.013(0.005)	0.028(0.003)	0.013(−0.001)	0.029(0.001)
β_{22} : 1	0.971(0.517)	0.268(0.515)	0.216(0.537)	0.168(0.433)	0.107(0.061)	0.044(0.014)
π_1 : 0.3	0.001(0.023)	0.001(0.025)	0.001(0.022)	0.003(0.025)	0.002(0.002)	0.001(0.002)
Case Ⅵ: 20% gross outliers						
β_{10} : 1	0.015(−0.011)	34.70(0.017)	18.39(0.038)	0.013(0.009)	35.96(−0.011)	0.019(0.004)
β_{20} : 1	1.0e+4(101.5)	775.6(0.018)	41.10(12.81)	257.1(−0.003)	123.1(−0.011)	0.007(−0.004)
β_{11} : −1	6.395(2.532)	16.17(0.069)	2.934(3.348)	0.374(−0.015)	10.38(2.623)	0.018(−0.008)
β_{21} : 3	0.014(0.015)	5.582(−0.049)	16.28(10.02)	0.005(0.002)	2.171(−1.274)	0.009(0.004)
β_{12} : 1	0.018(0.014)	1.929(0.015)	45.93(0.043)	0.015(0.012)	6.839(0.000)	0.021(0.012)
β_{22} : 1	0.013(−0.005)	8.953(−0.001)	118.8(30.27)	0.007(−0.006)	6.701(0.017)	0.046(−0.014)
π_1 : 0.3	0.250(0.500)	0.031(0.067)	0.026(0.472)	0.012(0.071)	0.038(0.200)	0.005(0.063)

Example 2.2: In this example, we consider a case when the number of components is larger than two and the components are close. We generate the independent and identically distributed (i.i.d.) data $\{(x_i, y_i), i = 1, \dots, n\}$ from the model

$$y_i = \begin{cases} 1+x_{i1}+\gamma_i\sigma+\epsilon_{i1}, & if\ z_{i1}=1 \\ 1+2x_{i1}+\gamma_i\sigma+\epsilon_{i2}, & if\ z_{i2}=2 \\ 1+3x_{i1}+\gamma_i\sigma+\epsilon_{i3}, & if\ z_{i3}=3 \end{cases}$$

where $P(z_{i1}=1)=P(z_{i2}=2)=0.3$, $P(z_{i3}=3)=0.4$, and $\sigma=1$. x_{i1} is independently generated from $N(0, 1)$. Note that in this example all three components have the same sign of the slopes and the three components are very close. We consider the following five cases for the component error densities:

Case Ⅰ: ϵ_1, ϵ_2, and ϵ_3 have the same distribution from $N(0, 1)$.

Case Ⅱ: $\epsilon_1 \sim t_9$, $\epsilon_2 \sim t_6$ and $\epsilon_3 \sim t_3$.

Case Ⅲ: $\epsilon_1 \sim N(0, 1)$, $\epsilon_2 \sim N(0, 1)$, and $\epsilon_3 \sim t_3$.

Case Ⅳ: ϵ_1, ϵ_2, and ϵ_3 have the same distribution from $N(0, 1)$ with 5% of low leverage outliers.

Case Ⅴ: ϵ_1, ϵ_2, and ϵ_3 have the same distribution from $N(0, 1)$ with 5% of high leverage outliers.

Case Ⅵ: ϵ_1, ϵ_2, and ϵ_3 have the same distribution from $N(0, 1)$ with 20% of gross outliers where outliers are concentrated in a very remote position.

Similar to example 2. 1, all γ_i's are set to be zero in Case Ⅰ, Case Ⅱ and Case Ⅲ. In Case Ⅳ and Case Ⅴ, 5% of nonzero mean-shift parameters, which are from the third component, are randomly generated from a uniform distribution between 11 and 13 and all other γ_i's are set to be zero. In Case Ⅴ, the 5% outliers are replaced by $x_{i1} = 20$ and $y_i = 200$, from which high leverage outliers are generated. In Case Ⅵ, 20% of nonzero mean-shift parameter is randomly generated from a uniform distribution between 101 and 102 for the second component, and 80% of γ_i's are set to be zero.

Table 2. 3 and Table 2. 4 report MSE (RB) of the parameter estimates for each estimation method for sample size $n = 200$ and $n = 400$, respectively. The findings are similar to Example 2. 1.

Table 2.3 ***MSE (RB) of point estimates for n=200 in Example 2.2***

TRUE	ML	Mixregt	MixregL	TLE	MEM-bisquare	CWRM
Case Ⅰ: $\epsilon_1 \sim N(0, 1)$, $\epsilon_2 \sim N(0, 1)$, and $\epsilon_3 \sim N(0, 1)$						
β_{10} : 1	0.092(0.092)	0.599(0.071)	0.395(0.021)	0.663(0.167)	0.537(0.129)	0.129(0.017)
β_{20} : 1	0.092(0.003)	0.640(0.091)	0.601(0.014)	0.598(0.155)	0.813(0.170)	0.114(0.004)
β_{30} : 1	0.041(0.017)	0.390(0.079)	0.324(0.003)	0.493(0.115)	0.415(0.064)	0.421(-0.012)
β_{11} : 1	0.023(0.005)	0.225(0.357)	0.171(0.148)	0.286(0.489)	0.153(0.048)	0.263(0.448)
β_{21} : 2	0.017(0.016)	0.120(0.279)	0.134(0.253)	0.106(0.261)	0.154(0.304)	0.054(-0.019)
β_{31} : 3	0.183(0.089)	1.044(1.028)	0.799(0.897)	1.087(1.055)	0.753(0.857)	0.980(0.985)
π_1 : 0.3	0.008(0.009)	0.009(0.015)	0.013(0.021)	0.007(0.002)	0.014(0.052)	0.010(0.035)
Case Ⅱ: $\epsilon_1 \sim t_9$, $\epsilon_2 \sim t_6$, and $\epsilon_3 \sim t_3$						
β_{10} : 1	14.66(0.006)	0.808(0.037)	0.390(0.038)	0.855(0.012)	1.028(0.010)	0.528(0.021)
β_{20} : 1	2.195(0.021)	0.168(0.014)	0.164(0.046)	0.092(0.017)	0.124(0.020)	0.115(0.013)
β_{30} : 1	10.63(0.076)	0.597(0.063)	0.352(0.076)	0.679(0.023)	0.867(0.038)	0.823(0.016)
β_{11} : 1	13.46(0.205)	0.266(0.302)	0.197(0.064)	0.278(0.405)	0.268(0.008)	0.121(0.231)
β_{21} : 2	0.365(0.245)	0.158(0.261)	0.183(0.291)	0.134(0.267)	0.216(0.274)	0.053(0.155)
β_{31} : 3	8.169(0.529)	0.956(0.990)	0.670(0.771)	0.989(1.024)	0.692(0.729)	0.800(0.951)
π_1 : 0.3	0.076(0.166)	0.012(0.027)	0.017(0.018)	0.009(0.031)	0.016(0.060)	0.010(0.048)
Case Ⅲ: $\epsilon_1 \sim N(0, 1)$, $\epsilon_2 \sim N(0, 1)$, and $\epsilon_3 \sim t_3$						
β_{10} : 1	18.13(0.015)	0.652(0.010)	0.295(0.063)	0.669(0.034)	0.631(0.015)	0.425(0.034)
β_{20} : 1	2.889(0.027)	0.712(0.184)	0.410(0.044)	0.708(0.218)	1.021(0.035)	0.015(0.004)
β_{30} : 1	16.56(0.003)	0.470(0.017)	0.375(0.027)	0.507(0.021)	0.578(0.035)	0.122(-0.009)
β_{11} : 1	78.58(0.159)	0.189(0.299)	0.142(0.049)	0.207(0.325)	0.173(0.017)	0.161(0.323)
β_{21} : 2	0.553(0.245)	0.162(0.293)	0.153(0.264)	0.169(0.292)	0.181(0.291)	0.065(-0.191)
β_{31} : 3	19.42(0.479)	0.929(0.981)	0.698(0.792)	0.935(0.965)	0.701(0.761)	0.885(-0.959)
π_1 : 0.3	0.068(0.115)	0.011(0.009)	0.015(0.038)	0.009(0.001)	0.022(0.031)	0.005(0.052)
Case Ⅳ: ϵ_1, ϵ_2, ϵ_3, $\sim N(0, 1)$ with 5% of low leverage outliers						
β_{10} : 1	17.02(0.103)	1.208(0.150)	0.352(0.502)	0.374(0.231)	11.17(0.209)	0.272(0.482)
β_{20} : 1	78.27(1.234)	1.305(0.951)	13.21(0.974)	1.354(0.903)	6.120(0.486)	0.683(0.822)
β_{30} : 1	74.59(1.205)	2.443(1.203)	0.945(0.946)	1.672(1.167)	9.435(0.960)	0.804(0.903)

Table 2.3 *MSE* (*RB*) *of point estimates for n=200 in Example* 2.2(Continued table)

TRUE	ML	Mixregt	MixregL	TLE	MEM-bisquare	CWRM
β_{11} : 1	0.161(0.127)	0.194(0.045)	0.197(0.353)	0.138(0.105)	6.923(0.211)	0.189(0.353)
β_{21} : 2	0.100(0.222)	0.111(0.183)	0.069(0.216)	0.125(0.231)	0.505(0.563)	0.033(−0.123)
β_{31} : 3	0.847(0.924)	0.897(0.948)	1.036(1.069)	0.941(0.970)	3.014(0.954)	0.983(−0.979)
π_1 : 0.3	0.022(0.036)	0.008(0.037)	0.012(0.003)	0.008(0.051)	0.032(0.124)	0.003(0.047)
			Case Ⅴ: ϵ_1, ϵ_2, ϵ_3, ~N (0, 1) with 5% of high leverage outliers			
β_{10} : 1	0.098(−0.037)	0.262(0.053)	0.125(0.033)	0.381(−0.211)	0.149(−0.073)	0.058(0.098)
β_{20} : 1	1.222(1.063)	1.434(1.203)	1.422(1.122)	1.282(0.710)	1.244(1.020)	1.253(1.064)
β_{30} : 1	15.29(0.850)	10.45(0.177)	10.85(0.659)	1.720(1.300)	7.065(1.109)	1.374(1.161)
β_{11} : 1	0.091(−0.001)	0.261(0.455)	0.165(0.226)	0.164(0.155)	0.140(−0.009)	0.073(0.136)
β_{21} : 2	0.026(0.016)	0.053(−0.095)	0.040(−0.051)	0.136(−0.257)	0.078(−0.112)	0.026(−0.055)
β_{31} : 3	67.87(6.907)	48.07(6.941)	48.03(6.917)	0.981(−1.000)	32.53(6.793)	0.257(−0.454)
π_1 : 0.3	0.009(0.037)	0.022(0.116)	0.022(0.086)	0.008(−0.390)	0.014(0.028)	0.012(0.095)
			Case Ⅵ: 20% gross outliers			
β_{10} : 1	8.937(0.175)	317.9(0.227)	0.026(0.078)	373.7(−0.011)	0.397(0.629)	0.111(0.287)
β_{20} : 1	4525(1.195)	892.7(0.765)	472.5(1.429)	420.5(0.774)	52.96(0.630)	0.602(0.768)
β_{30} : 1	5787(102.3)	1069(0.945)	577.7(102.3)	735.8(1.186)	0.404(0.632)	0.797(0.883)
β_{11} : 1	0.171(0.172)	68.57(0.077)	0.011(0.041)	18.76(−0.012)	0.397(0.618)	0.049(0.124)
β_{21} : 2	0.076(−0.182)	0.207(−0.308)	0.005(−0.087)	0.253(−0.363)	0.158(−0.379)	0.063(−0.167)
β_{31} : 3	0.914(−0.937)	26.23(−0.934)	0.082(−0.913)	0.841(−0.907)	1.906(−1.378)	0.975(−0.987)
π_1 : 0.3	0.011(0.062)	0.011(0.021)	0.001(0.052)	0.020(−0.017)	0.001(0.033)	0.007(0.064)

Table 2.4 *MSE* (*RB*) *of point estimates for n =400 in Example* 2.2

TRUE	MLE	Mixregt	MixregL	TLE	MEM-bisquare	CWRM
			Case Ⅰ: ϵ_1~N (0, 1), ε_2~N (0, 1), and ε_3~N (0, 1)			
β_{10} : 1	0.038(0.005)	0.064(0.021)	0.056(0.020)	0.070(0.031)	0.043(0.080)	0.085(0.010)
β_{20} : 1	0.064(0.023)	0.056(0.064)	0.041(0.038)	0.054(0.004)	0.072(0.021)	0.068(−0.014)
β_{30} : 1	0.026(0.007)	0.047(0.113)	0.045(0.081)	0.049(0.011)	0.275(0.045)	0.051(−0.014)
β_{11} : 1	0.015(0.007)	0.275(0.052)	0.163(0.155)	0.033(0.058)	0.116(0.017)	0.237(0.448)

Table 2.4 ***MSE (RB) of point estimates for n =400 in Example* 2.2(Continued table)**

TRUE	MLE	Mixregt	MixregL	TLE	MEM-bisquare	CWRM
β_{21} : 2	0.018(0.030)	0.095(0.247)	0.107(0.100)	0.094(0.236)	0.159(0.298)	0.041(0.163)
β_{31} : 3	0.827(0.094)	1.140(1.082)	0.985(0.891)	1.152(1.093)	0.827(0.919)	1.015(0.998)
π_1 : 0.3	0.007(0.008)	0.008(0.013)	0.010(0.012)	0.006(0.009)	0.014(0.068)	0.005(0.035)
Case Ⅱ : $\epsilon_1 \sim t_9$, $\epsilon_2 \sim t_6$, and $\epsilon_3 \sim t_3$						
β_{10} : 1	58.11(0.001)	0.685(0.052)	0.208(0.017)	0.676(0.040)	0.527(0.032)	0.617(0.011)
β_{20} : 1	30867(0.001)	0.680(0.132)	0.362(0.013)	0.826(0.036)	1.511(0.044)	0.108(0.009)
β_{30} : 1	26.91(0.024)	0.421(0.061)	0.184(0.030)	0.413(0.033)	0.574(0.027)	0.012(0.011)
β_{11} : 1	136.9(0.131)	0.225(0.341)	0.103(0.002)	0.201(0.297)	0.125(0.022)	0.104(0.249)
β_{21} : 2	0.426(0.240)	0.151(0.285)	0.174(0.285)	0.160(0.302)	0.195(0.301)	0.039(0.126)
β_{31} : 3	20.12(0.177)	1.047(1.025)	0.735(0.868)	1.050(1.017)	0.758(0.893)	0.863(0.987)
π_1 : 0.3	0.094(0.148)	0.011(0.019)	0.014(0.031)	0.008(0.023)	0.016(0.036)	0.003(0.044)
Case Ⅲ : $\epsilon_1 \sim N(0, 1)$, $\epsilon_2 \sim N(0, 1)$, and $\epsilon_3 \sim t_3$						
β_{10} : 1	19.10(0.008)	0.679(0.039)	0.325(0.019)	0.627(0.060)	0.331(0.023)	0.725(0.016)
β_{20} : 1	5.947(0.005)	0.640(0.172)	0.458(0.140)	0.720(0.169)	0.760(0.026)	0.012(0.005)
β_{30} : 1	13.48(0.016)	0.435(0.077)	0.236(0.165)	0.407(0.085)	0.686(0.012)	0.913(0.012)
β_{11} : 1	14.19(0.170)	0.226(0.426)	0.125(0.146)	0.212(0.366)	0.129(0.078)	0.145(0.341)
β_{21} : 2	10233(0.229)	0.118(0.268)	0.121(0.290)	0.131(0.281)	0.159(0.237)	0.042(0.146)
β_{31} : 3	15.59(0.374)	1.130(1.093)	0.729(0.860)	1.099(1.027)	0.757(0.866)	0.967(0.989)
π_1 : 0.3	0.076(0.176)	0.008(0.017)	0.006(0.003)	0.006(0.024)	0.019(0.032)	0.005(0.048)
Case Ⅳ : ϵ_1, ϵ_2, ϵ_3, $\sim N(0, 1)$ with 5% of low leverage outliers						
β_{10} : 1	5.250(0.098)	0.286(0.260)	6.355(0.229)	0.264(0.324)	7.402(0.112)	0.242(0.467)
β_{20} : 1	69.39(1.120)	1.118(0.709)	16.87(0.850)	1.201(0.662)	10.36(0.302)	0.678(0.818)
β_{30} : 1	95.49(12.63)	1.689(1.318)	7.742(1.262)	1.786(1.330)	11.04(0.913)	0.778(0.875)
β_{11} : 1	0.125(0.080)	0.138(0.228)	0.083(0.017)	0.150(0.226)	5.432(0.100)	0.169(0.368)
β_{21} : 2	0.068(0.129)	0.111(0.225)	0.059(0.157)	0.125(0.244)	0.584(0.750)	0.019(0.093)
β_{31} : 3	0.897(0.930)	1.057(1.043)	0.810(0.846)	1.070(1.048)	2.181(1.038)	1.004(1.001)
π_1 : 0.3	0.013(0.300)	0.007(0.046)	0.015(0.074)	0.007(0.062)	0.025(0.112)	0.003(0.041)

Table 2.4 ***MSE*** **(*****RB*****)** ***of point estimates for n*** **=400** ***in Example*** **2.2(Continued table)**

TRUE	MLE	Mixregt	MixregL	TLE	MEM-bisquare	CWRM
			Case V: ϵ_1, ϵ_2, ϵ_3, $\sim N(0, 1)$ with 5% of high leverage outliers			
β_{10} : 1	0.084(0.001)	0.343(0.027)	0.295(−0.035)	0.531(0.047)	0.093(0.025)	0.018(−0.009)
β_{20} : 1	0.044(−0.020)	0.300(−0.055)	0.240(−0.045)	0.695(0.046)	0.070(−0.032)	0.009(−0.005)
β_{30} : 1	9.697(0.284)	8.753(0.511)	10.36(0.544)	0.400(−0.045)	4.253(0.094)	0.013(0.033)
β_{11} : 1	0.076(−0.036)	0.323(0.584)	0.246(0.524)	0.191(0.278)	0.080(0.004)	0.142(0.326)
β_{21} : 2	0.029(0.068)	0.069(−0.195)	0.060(−0.137)	0.113(−0.262)	0.065(0.043)	0.037(−0.127)
β_{31} : 3	78.08(6.936)	48.20(6.924)	48.07(6.923)	1.030(−1.025)	38.11(6.891)	0.955(−0.977)
π_1 : 0.3	0.017(0.047)	0.031(0.144)	0.027(0.124)	0.010(−0.045)	0.018(0.051)	0.004(0.046)
			Case VI: 20% gross outliers			
β_{10} : 1	0.112(0.089)	111.4(0.177)	0.009(−0.006)	126.9(−0.129)	0.402(0.632)	0.085(0.281)
β_{20} : 1	4991(1.244)	630.5(0.751)	368.1(102.44)	524.4(0.895)	0.402(0.632)	0.639(0.791)
β_{30} : 1	5519(102.3)	1320(0.985)	314.8(1.175)	1001(1.186)	0.403(0.632)	0.816(0.899)
β_{11} : 1	0.113(0.064)	23.89(0.114)	0.017(0.085)	21.35(0.010)	0.415(0.634)	0.030(0.120)
β_{21} : 2	0.040(−0.092)	0.121(−0.235)	0.003(−0.126)	0.201(−0.267)	0.136(−0.366)	0.047(−0.147)
β_{31} : 3	0.915(−0.954)	12.97(−0.954)	0.058(−0.867)	13.64(−0.904)	1.856(−1.365)	1.006(−1.014)
π_1 : 0.3	0.007(0.036)	0.008(0.021)	0.001(0.018)	0.017(−0.003)	0.002(0.033)	0.007(0.071)

2.4 Discussion

In this chapter, we reviewed some popular robust estimation methods for mixture of regression and compared some of them with the traditional MLE using simulation studies. The results show that the normality based MLE failed when there are outliers or the error has the heavy-tailed distribution while all robust methods work reasonably well. The trimmed cluster weighted restricted model

(CWRM) and the trimmed likelihood estimate (TLE) work well if the trimming proportion is properly chosen. When the proportion of outliers is large, say 20%, Mixregt, MixregL and MEM-bisquare will break down.

It is also interesting to investigate the sample breakdown points for the reviewed robust methods. We provide an empirical idea of "breakdown points" for the compared procedures in very extreme cases. However, we should note that the analysis of breakdown point for traditional linear regression can not be directly applied to mixture regression. For example, the breakdown point of TLE for traditional linear regression does not apply to the mixture regression. García-Escudero et al. (2010) also stated that the traditional definition of break-down point is not the right one to quantify the robustness of clustering regression procedures to outliers, since the robustness of these procedures is not only data dependent but also cluster dependent.

Chapter 3

Outlier Detection and Robust Mixture Modeling Using Nonconvex Penalized Likelihood

3.1 Introduction

Nowadays finite mixture distributions are increasingly important in modeling a variety of random phenomena (Everitt and Hand, 1981; Titterington et al., 1985; McLachlan and Basford, 1988; Lindsay, 1995; Böhning, 1999). The m-component finite normal mixture distribution has probability density

$$f(y;\boldsymbol{\theta})=\sum_{j=1}^{m}\pi_j\phi(y;\mu_j,\sigma_j^2) \tag{3-1}$$

where $\boldsymbol{\theta}=(\pi_1, \mu_1, \sigma_1; \dots ; \pi_m, \mu_m, \sigma_m)^T$ collects all the unknown parameters, $\phi(\cdot\ ;\mu, \sigma^2)$ denotes the density function of $N(\mu, \sigma^2)$, and π_j is the proportion of the jth subpopulation $\sum_{j=1}^{m}\pi_j=1$. Given observations $(y_1, \dots,$

y_n) from model (3-1), the maximum likeli-hood estimator (MLE) of $\boldsymbol{\theta}$ is given by,

$$\hat{\boldsymbol{\theta}}_{\text{MLE}} = \arg\max_{\boldsymbol{\theta}} \sum_{i=1}^{n} \log\left\{\sum_{j=1}^{m} \pi_j \phi(y_i; \mu_j, \sigma_j^2)\right\} \tag{3-2}$$

which does not have an explicit form and is usually calculated by the EM algorithm (Dempster et al., 1977).

The MLE based on the normality assumption possesses many desirable properties such as asymptotic efficiency, however, it is sensitive to the presence of outliers. For the estimation of a single location, many robust methods have been proposed, including the M-estimator (Huber, 1981), the Least Median of Squares (LMS) estimator (Siegel, 1982), the least trimmed squares (LTS) estimator (Rousseeuw, 1983), the S-estimates (Rousseeuw and Yohai, 1984), the MM-estimator Yohai (1987), and the weighted least squares estimator (REWLSE) (Gervini and Yohai, 2002). In contrast, there is much less research on robust estimation of the mixture model, in part because it is not straightforward to replace the log-likelihood in (3-2) by a robust criterion similar to the M-estimation. Peel and McLachlan (2000) proposed a robust mixture modeling using t distribution. Markatou (2000) and Qin et al. (2013) proposed using a weighted likelihood for each data point to robustify the estimation procedure for mixture models. Fujisawa and Eguchi (2005) proposed a robust estimation method in normal mixture model using a modified likelihood function. Neykov et al. (2007) proposed robust fitting of mixtures using the trimmed likelihood. Other related robust methods on mixture models include Henning (2002, 2003), Shen et al. (2004), Bai et al. (2012) and Bashir and Carter (2012).

In Chapter 3, we propose a new robust mixture modeling approach via a mean-shift penalization, which achieves simultaneous outlier detection and robust parameter estimation. A case-specific mean shift parameter vector is added to the mean structure of the mixture model, and it is assumed to be sparse for capturing the rare but possibly severe outlying effects induced by the potential outliers. When the mixture components are assumed to have equal variances, our method directly extends the robust linear regression approaches proposed by She and Owen (2011) and Lee et al. (2012). However, even in this case the optimization of the penalized mixture log-likelihood is not trivial, especially for the SCAD penalty (Fan and Li, 2001). For the general case of unequal component variances, the variance heterogeneity of different components complicates the declaration and detection of the outliers, and the naive mean-shift model for the equal variance case is no longer appropriate. We thus propose a scale-free and case-specific mean-shift formulation to achieve the robustness in the general mixture model setup.

The rest of Chapter 3 is organized as follows. In Section 3.2, we introduce the robust mixture model via mean shift penalization (RMM) and propose an iterative thresholding embedded EM algorithm for the cases of both equal variances and unequal variances. In addition, we also propose a data adaptive tuning parameter selection method. In Section 3.3, we compare the proposed methods to several existing methods via simulation studies. A real application is presented in Section 3.4. Section 3.5 contains a discussion of related work.

3.2 Robust Mixture Model via Mean-Shift Penalization

In this section, we will introduce the proposed robust mixture modeling approach via mean-shift penalization (RMM). To focus on the main idea, we restrict our attention on the normal mixture model. The proposed approach can be readily extended to other mixture models, such as gamma mixture, poisson mixture, and logistic mixture. Due to the inherent difference between the case of equal component variances and the case of unequal component variances, we shall discuss them separately.

3.2.1 RMM for Equal Component Variances

Assume the mixture components have equal variances, i.e., $\sigma_1^2 = \ldots = \sigma_m^2 = \sigma^2$. The proposed robust mixture model with a mean-shift parameterization is to assume that the observations $(y_1, \ldots, y_n)$ come from the following mixture density

$$f(y_i; \boldsymbol{\theta}, \gamma_i) = \sum_{j=1}^{m} \pi_j \phi(y_i - \gamma_i; \mu_j, \sigma^2), \qquad i = 1, \ldots, n \qquad (3-3)$$

where $\boldsymbol{\theta} = (\pi_1, \mu_1, \ldots, \pi_m, \mu_m, \sigma)^T$ and γ_i is the mean shift parameter for the ith observation, which is nonzero when the ith observation is an outlier and is zero otherwise. Therefore, the sparse estimation of γ_i provides a direct way to identify and accommodate outliers.

Due to the sparsity assumption of γ_i, we propose to maximize the following penalized log-likelihood criterion to conduct model estimation and outlier detection,

$$pl_1(\boldsymbol{\theta},\gamma)=l_1(\boldsymbol{\theta},\gamma)-\sum_{i=1}^{n}\frac{1}{w_i}P_\lambda(|\gamma_i|) \quad (3-4)$$

where $l_1(\boldsymbol{\theta}, \gamma)=\sum_{i=1}^{n}\log\{\sum_{j=1}^{m}\pi_j\phi(y_i-\gamma_i;\mu_j,\sigma^2)\}$, $\gamma=(\gamma_1, \ldots, \gamma_n)$, w_is are the weights to reflect the prior information about how likely it is that the y_is are outliers, $P_\lambda(\cdot)$ is some penalty function used to induce the sparsity in $\boldsymbol{\gamma}$, and $\boldsymbol{\lambda}$ is a tuning parameter controlling the number of outliers, i.e., the number of nonzero γ_i. To focus on the key idea, we mainly consider $w_1=w_2=\ldots=w_n=w$ and discuss the choice of w for different penalty functions.

The commonly used penalty functions include the ℓ_1 norm penalty (Donoho and Johnstone, 1994; Tibshirani, 1996, 1997) $P_\lambda(\gamma)=\lambda|\gamma|$, the ℓ_0 penalty (Antoniadis, 1997)

$$P_\lambda(\gamma)=\frac{\lambda^2}{2}I(\gamma\neq 0) \quad (3-5)$$

and the SCAD penalty (Fan and Li, 2001)

$$P_\lambda(\gamma)=\begin{cases}\lambda|\gamma|, & \text{if } |\gamma|\leqslant\lambda\\ -\left(\frac{\gamma^2-2a\lambda|\gamma|+\lambda^2}{2(a-1)}\right), & \text{if } \lambda<|\gamma|\leqslant a\lambda\\ \frac{(a+1)\lambda^2}{2}, & \text{if } |\gamma|>a\lambda\end{cases} \quad (3-6)$$

where a is a constant usually set to be 3.7. In penalized estimation, each of the above penalty forms corresponds to a thresholding rule, e.g., ℓ_1 penalization corresponds to a soft-threshing rule and ℓ_0 penalization corresponds to a hard-thresholding rule. We mainly focus on the nonconvex hard penalty and SCAD penalty, due to their superior performance in sparse estimation.

We propose a thresholding embedded EM algorithm to maximize the objective function (3-4). Define the latent variable z_{ij} and let $\mathrm{z}_i=(z_{i1}, \dots, z_{im})$. The complete penalized log-likelihood function based on the complete data $\{(y_i, \mathbf{z}_i), i=1, 2, \dots, n\}$ is

$$pl_1^c(\boldsymbol{\theta},\boldsymbol{\gamma})=\sum_{i=1}^{n}\sum_{j=1}^{m} z_{ij}\log\left\{\pi_j\phi(y_i-\gamma_i;\mu_j,\sigma^2)\right\}-\sum_{i=1}^{n}\frac{1}{w}P_\lambda(|\gamma_i|) \tag{3-7}$$

Based on the construction of the EM algorithm, in the E-step, given the current estimate $\boldsymbol{\theta}^{(k)}$ and $\boldsymbol{\gamma}^{(k)}$ at the kth iteration, we need to find the condition expectation of the complete penalized log-likelihood function (3-7), i.e., $E\{pl_1^c(\boldsymbol{\theta}, \boldsymbol{\gamma}) \mid \boldsymbol{\theta}^{(k)}, \boldsymbol{\gamma}^{(k)}\}$, which simplifies to the calculation of $E(z_{ij} \mid \mathrm{y_i}; \boldsymbol{\theta}^{(k)}, \boldsymbol{\gamma}^{(k)})$:

$$p_{ij}^{(k+1)}=\mathrm{E}(z_{ij}|y_i;\boldsymbol{\theta}^{(k)},\boldsymbol{\gamma}^{(k)})=\frac{\pi_j^{(k)}\phi(y_i-\gamma_i^{(k)};\mu_j^{(k)},\sigma^{2(k)})}{\sum_{j=1}^{m}\pi_j^{(k)}\phi(y_i-\gamma_i^{(k)};\mu_j^{(k)},\sigma^{2(k)})}$$

In the M-step, we then update $(\boldsymbol{\theta}, \boldsymbol{\gamma})$ by maximizing $E\{pl_1^c(\boldsymbol{\theta}, \boldsymbol{\gamma}) \mid \boldsymbol{\theta}^{(k)}, \boldsymbol{\gamma}^{(k)}\}$. There is no explicit solution, except for the $\pi_j s$: $\pi_j^{(k+1)}=\sum_{i=1}^{n} p_{ij}^{(k+1)}/n$. We propose to iterate the following two steps until convergence to get $\{\mu_j^{(k+1)}, j=1, \dots, m, \sigma^{(k+1)}, \boldsymbol{\gamma}^{(k+1)}\}$:

Step 1: Given μ_j s and σ update $\boldsymbol{\gamma}$ by maximizing

$$\sum_{i=1}^{n}\sum_{j=1}^{m}p_{ij}^{(k+1)}\log\phi(y_i-\gamma_i;\mu_j,\sigma^2)-\sum_{i=1}^{n}\frac{1}{w}P_\lambda(|\gamma_i|)$$

which is equivalently to minimizing

$$\frac{1}{2}\left\{\gamma_i - \sum_{j=1}^{m} p_{ij}^{(k+1)}(y_i - \mu_j)\right\}^2 + \frac{1}{w}\sigma^2 P_\lambda(|\gamma_i|) \tag{3-8}$$

separately for each γ_i.

Step 2: Given $\boldsymbol{\gamma}$, the μ_j s and σ are updated by

$$\mu_j \leftarrow \frac{\sum_{i=1}^n p_{ij}^{(k+1)}(y_i - \gamma_i)}{\sum_{i=1}^n p_{ij}^{(k+1)}}, j = 1, \ldots, m$$

$$\sigma^2 \leftarrow \frac{\sum_{j=1}^m \sum_{i=1}^n p_{ij}^{(k+1)}(y_i - \gamma_i - \mu_j)^2}{n}$$

Note that for the hard penalty, $w^{-1}\sigma^2 P_\lambda(|\gamma_i|) = \sigma P_{\lambda*}(|\gamma_i|)$, where $\lambda^* = \frac{\sigma}{\sqrt{w}}\lambda$. Therefore, if λ is chosen data adaptively, we can simply set $w=1$ for the hard penalty. However, for the SCAD penalty, such property does not hold and the solution may be affected nonlinearly by the ratio σ^2/w. In order to mimic the unscaled SCAD and use the same a value as suggested by Fan and Li (2001), we need to make sure σ^2/w is close to 1. Therefore, we propose to set $w=\hat{\sigma}^2$ for SCAD penalty, where $\hat{\sigma}^2$ is a robust estimate of σ^2 such as the estimate from the trimmed likelihood estimation (Neykov et al., 2007) or the estimator using the hard penalty assuming $w=1$.

If the hard penalty is used, (3-8) is minimized by the hard thresholding rule. However, if the SCAD penalty is used, we prove in the following proposition that the minimizer of (3-8) is given by a modified SCAD thresholding rule.

Proposition 3.1: *Let*

$$\xi_i = \sum_{j=1}^{m} p_{ij}^{(k+1)}(y_i - \mu_j) \tag{3-9}$$

If the penalty function in (3-8) is the hard penalty (3-5), then the thresholding rule to minimize (3-8) is

$$\hat{\gamma}_i = \Theta_{hard}(\xi_i; \lambda, \sigma) = \begin{cases} 0, & \text{if } |\xi_i| \leqslant \sigma\lambda \\ \xi_i, & \text{if } |\xi_i| > \sigma\lambda \end{cases}$$

If the penalty function in (3-8) is the SCAD penalty (3-6), then the thresholding rule to minimize (3-8) is

(1) *when* $\sigma^2/\hat{\sigma}^2 < a-1$,

$$\hat{\gamma}_i = \Theta_{SCAD}(\xi_i; \lambda, \sigma) = \begin{cases} sgn(\xi_i)\left(|\xi_i| - \frac{\sigma^2\lambda}{\hat{\sigma}^2}\right)_+, & \text{if } |\xi_i| \leqslant \lambda + \frac{\sigma^2\lambda}{\hat{\sigma}^2} \\ \frac{\frac{\hat{\sigma}^2}{\sigma^2}(a-1)\xi_i - sgn(\xi_i)a\lambda}{\frac{\hat{\sigma}^2}{\sigma^2}(a-1)-1}, & \text{if } \lambda + \frac{\sigma^2\lambda}{\hat{\sigma}^2} < |\xi_i| \leqslant a\lambda \\ \xi_i, & \text{if } |\xi_i| > a\lambda \end{cases} \tag{3-10}$$

(2) *when* $a-1 \leqslant \sigma^2/\hat{\sigma}^2 \leqslant a+1$,

$$\hat{\gamma}_i = \Theta_{SCAD}(\xi_i; \lambda, \sigma) = \begin{cases} sgn(\xi_i)\left(|\xi_i| - \frac{\sigma^2\lambda}{\hat{\sigma}^2}\right)_+, & \text{if } |\xi_i| \leqslant \frac{a+1+\frac{\sigma^2}{\hat{\sigma}^2}}{2}\lambda \\ \xi_i, & \text{if } |\xi_i| > \frac{a+1+\frac{\sigma^2}{\hat{\sigma}^2}}{2}\lambda \end{cases} \tag{3-11}$$

(3) *when* $\sigma^2/\hat{\sigma}^2 > a+1$,

$$\hat{\gamma}_i = \Theta_{SCAD}(\xi_i; \lambda, \sigma) = \begin{cases} 0, & \text{if } |\xi_i| \leqslant \sqrt{\frac{\sigma^2(a+1)}{\hat{\sigma}^2}}\lambda \\ \xi_i, & \text{if } |\xi_i| > \sqrt{\frac{\sigma^2(a+1)}{\hat{\sigma}^2}}\lambda \end{cases} \tag{3-12}$$

The detailed EM algorithm to maximize the penalized log-likelihood (3-4) is summarized in Algorithm 3. 1. The convergence property of the proposed algorithm is summarized in Theorem 3. 1 below, which follows directly from the property of the EM algorithm, and hence its proof is omitted.

Theorem 3. 1: Each iteration of E step and M step of Algorithm 1 monotonically non-decreases the penalized log-likelihood (3. 4), i.e., $pl_1(\boldsymbol{\theta}^{(k+1)}, \boldsymbol{\gamma}^{(k+1)}) \geqslant pl_1(\boldsymbol{\theta}^{(k)}, \boldsymbol{\gamma}^{(k)})$, for all $k \geqslant 0$.

Given the initial parameter estimate $\boldsymbol{\theta}^{(0)}$, at $(k+1)th$ step, the EM algorithm iterates the following E-step and M-step:

Algorithm 3. 1: *Initialize* $\boldsymbol{\theta}^{(0)}$ and $\gamma^{(0)}$. Set $k \leftarrow 0$.

E-Step: *Compute the classification probabilities*

$$p_{ij}^{(k+1)} = E(z_{ij}|y_i;\boldsymbol{\theta}^{(k)}) = \frac{\pi_j^{(k)}\phi(y_i-\gamma_i^{(k)};\mu_j^{(k)},\sigma^{2(k)})}{\sum_{j=1}^{m}\pi_j^{(k)}\phi(y_i-\gamma_i^{(k)};\mu_j^{(k)},\sigma^{2(k)})}$$

M-Step: Update $(\boldsymbol{\theta}, \boldsymbol{\gamma})$ by the following two steps:

(1) $\pi_j^{(k+1)} = \frac{\sum_{i=1}^{n} p_{ij}^{(k+1)}}{n}, j = 1, \ldots, m$

(2) *Iterating the following steps until convergence to obtain*

$\{\mu_j^{(k+1)}, j = 1, \ldots, m; \sigma^{2(k+1)}, \gamma^{(k+1)}\}$:

$$\gamma_i \leftarrow \Theta(\xi_i;\lambda,\sigma), i = 1,\ldots,n, \ \ where\ \xi_i = \sum_{j=1}^{m} p_{ij}^{(k+1)}(y_i-\mu_j)$$

$$\mu_j \leftarrow \frac{\sum_{i=1}^{n} p_{ij}^{(k+1)}(y_i-\gamma_i)}{\sum_{i=1}^{n} p_{ij}^{(k+1)}}, j = 1,\ldots,m$$

$$\sigma^2 \leftarrow \frac{\sum_{j=1}^{m}\sum_{i=1}^{n} p_{ij}^{(k+1)}(y_i-\gamma_i-\mu_j)^2}{n}$$

3.2.2 RMM for Unequal Component Variances

When the component variances are unequal, the naive mean shift model (3-3) can not be directly applied, due to the scale difference in the mixture components. To illustrate further, suppose the standard deviation in the first component is 1 and the standard deviation in the second component is 4. If some weighted residual ξ_i, defined in (3-9), equals to 5, then the ith observation is considered as an outlier if it is from the first component but should not be regarded as an outlier if it belongs to the second component. This suggests that the declaration of outliers in a mixture model shall take into account both the centers and the variabilities of all the components, i.e., an observation is considered as an outlier in the mixture model only if it is far away from all the component centers judged by their own component variabilities.

We propose the following scale-free mean shift model to incorporate the information on component variability,

$$f(y_i;\boldsymbol{\theta},\gamma_i)=\sum_{j=1}^{m}\pi_j\phi(y_i-\gamma_i\sigma_j;\mu_j,\sigma_j^2),\qquad i=1,\dots,n \tag{3-13}$$

where with some abuse of notation, $\boldsymbol{\theta}$ is redefined as $\boldsymbol{\theta}=(\pi_1,\mu_1,\sigma_1,\dots,\pi_m,\mu_m,\sigma_m)^T$. Given observations $(y_1, y_2, \dots, y_n)$, we estimate the parameters $\boldsymbol{\theta}$ and $\boldsymbol{\gamma}$ by maximizing the following penalized log-likelihood function:

$$pl_2(\boldsymbol{\theta},\boldsymbol{\gamma})=l_2(\boldsymbol{\theta},\boldsymbol{\gamma})-\sum_{i=1}^{n}\frac{1}{w_i}P_\lambda(|\gamma_i|) \tag{3-14}$$

where $l_2(\boldsymbol{\theta},\boldsymbol{\gamma})=\sum_{i=1}^{n}\log\{\sum_{j=1}^{m}\pi_j\phi(y_i-\gamma_i\sigma_j;\mu_j,\sigma_j^2)\}$. Since the

γ_is in (3-14) are scale free, for simplicity we set $w_1 = w_2 = \ldots = w_n = 1$ when no prior information is available.

We again propose a thresholding embedded EM algorithm to maximize (3-14). The complete penalized log-likelihood function constructed based on the complete data $\{(z_i, y_i), i = 1, 2, \ldots, n\}$, with the same setting of the binary label z_{ij} as the equal component variances case, is

$$pl_2^c(\boldsymbol{\theta}, \boldsymbol{\gamma}) = \sum_{i=1}^{n}\sum_{j=1}^{m} z_{ij} \log\left\{\pi_j \phi(y_i - \gamma_i \sigma_j; \mu_j, \sigma_j^2)\right\} - \sum_{i=1}^{n} P_\lambda(|\gamma_i|) \tag{3-15}$$

Similar to the arguments in Section 3.2.1, in the E step of the $(k+1)$th iteration, we only need to compute $E\{pl_2^c(\boldsymbol{\theta}, \boldsymbol{\gamma}) \mid \boldsymbol{\theta}^{(k)}, \boldsymbol{\gamma}^{(k)}\}$, which simplifies to the calculation of

$$p_{ij}^{(k+1)} = \mathrm{E}(z_{ij} | y_i; \boldsymbol{\theta}^{(k)}, \boldsymbol{\gamma}^{(k)}) = \frac{\pi_j^{(k)} \phi(y_i - \gamma_i^{(k)} \sigma_j^{(k)}; \mu_j^{(k)}, \sigma_j^{2(k)})}{\sum_{j=1}^{m} \pi_j^{(k)} \phi(y_i - \gamma_i^{(k)} \sigma_j^{(k)}; \mu_j^{(k)}, \sigma_j^{2(k)})}$$

In the M-step, we need to update $(\boldsymbol{\theta}, \boldsymbol{\gamma})$ by maximizing $E\{pl_2^c(\boldsymbol{\theta}, \boldsymbol{\gamma}) \mid \boldsymbol{\theta}^{(k)}, \boldsymbol{\gamma}^{(k)}\}$. Therefore, $\pi_j^{(k+1)} = \sum_{i=1}^{n} p_{ij}^{(k+1)}/n$, and $\{\mu_j^{(k+1)}, j = 1, \ldots, m, \sigma_j^{(k+1)}, \boldsymbol{\gamma}^{(k+1)}\}$ can be found by iterating the following three steps:

Step 1: Given $\boldsymbol{\gamma}$ and σ_j s, μ_j s are updated by

$$\mu_j \leftarrow \frac{\sum_{i=1}^{n} p_{ij}^{(k+1)} (y_i - \gamma_i \sigma_j)}{\sum_{i=1}^{n} p_{ij}^{(k+1)}}, j = 1, \ldots, m$$

Step 2: Given $\boldsymbol{\gamma}$ and μ_j s, σ_j s are updated by

$$\sigma_j^2 \leftarrow \arg\max_{\sigma_j} \sum_{i=1}^{n} p_{ij}^{(k+1)} \log \phi(y_i - \gamma_i \sigma_j; \mu_j, \sigma_j^2), j = 1, \ldots, m \tag{3-16}$$

Step 3: Given μ_j s and σ_j s, update $\boldsymbol{\gamma}$ by minimizing

$$\frac{1}{2}\left[\left\{\gamma_i-\sum_{j=1}^{m}\frac{p_{ij}^{(k+1)}}{\sigma_j}(y_i-\mu_j)\right\}^2\right]+P_\lambda(|\gamma_i|) \tag{3-17}$$

separately for each γ_i.

Note that, unlike the equal variances case, the update of σ_j^2 in (3-16) does not have explicit solution and requires some one-dimensional numerical algorithm to sovle, e.g., the Newton-Raphson method.

To minimize (3-17), we have the following thresholding solutions for using the hard and SCAD penalties, respectively:

$$\hat{\gamma}_i=\Theta_{hard}^*(\xi_i;\lambda)=\begin{cases}0, & \text{if } |\xi_i|\leqslant\lambda\\ \xi_i, & \text{if } |\xi_i|>\lambda\end{cases}$$

$$\hat{\gamma}_i=\Theta_{SCAD}^*(\xi_i;\lambda)=\begin{cases}sgn(\xi_i)(|\xi_i|-\lambda)_+, & \text{if } |\xi_i|\leqslant 2\lambda\\ \frac{(a-1)\xi_i-sgn(\xi_i)a\lambda}{a-2}, & \text{if } 2\lambda<|\xi_i|\leqslant a\lambda\\ \xi_i, & \text{if } |\xi_i|>a\lambda\end{cases}$$

where

$$\xi_i=\sum_{j=1}^{m}\frac{p_{ij}^{(k+1)}}{\sigma_j}(y_i-\mu_j)$$

The detailed thresholding embeded EM algorithm to maximize (3-14) can be summarized in Algorithm 3.2, with its convergence property summarized in Theorem 3.2.

Theorem 3.2: Each iteration of E step and M step of Algorithm 2 monotonically non-decreases the corresponding objective function, i.e., $pl_2(\boldsymbol{\theta}^{(k+1)}, \boldsymbol{\gamma}^{(k+1)}) \geqslant pl_2(\boldsymbol{\theta}^{(k)}, \boldsymbol{\gamma}^{(k)})$, for all $k \geqslant 0$.

Algorithm 3.2: *Initialize* $\boldsymbol{\theta}^{(0)}$ *and* $\boldsymbol{\gamma}^{(0)}$. *Set* $k \leftarrow 0$.

E-Step: *Compute the classification probabilities*

$$p_{ij}^{(k+1)} = E(z_{ij}|y_i; \boldsymbol{\theta}^{(k)}) = \frac{\pi_j^{(k)}\phi(y_i - \gamma_i^{(k)}\sigma_j^{(k)}; \mu_j^{(k)}, \sigma_j^{2(k)})}{\sum_{j=1}^m \pi_j^{(k)}\phi(y_i - \gamma_i^{(k)}\sigma_j^{(k)}; \mu_j^{(k)}, \sigma_j^{2(k)})}$$

M-Step: *Update* $(\boldsymbol{\theta}, \boldsymbol{\gamma})$ *by the following two steps*:

(1) $\pi_j^{(k+1)} = \frac{\sum_{i=1}^n p_{ij}^{(k+1)}}{n}, j = 1, \ldots, m.$

(2) *Iterating the following steps until convergence to obtain* $\{\mu_j^{(k+1)}, \sigma_j^{2(k+1)}, j=1, \ldots, m, \boldsymbol{\gamma}^{(k+1)}\}$:

$$\gamma_i \leftarrow \Theta^*(\xi_i; \lambda), \quad where\ \xi_i = \sum_{j=1}^m p_{ij}^{(k+1)}(y_i - \mu_j)/\sigma_j$$

$$\mu_j \leftarrow \frac{\sum_{i=1}^n p_{ij}^{(k+1)}(y_i - \gamma_i\sigma_j)}{\sum_{i=1}^n p_{ij}^{(k+1)}}$$

$$\sigma_j^2 \leftarrow \arg\max_{\sigma_j} \sum_{i=1}^n p_{ij}^{(k+1)} \log\phi(y_i - \gamma_i\sigma_j; \mu_j, \sigma_j^2)$$

3.2.3 Tuning Parameter Selection

In order to apply (3-4) and (3-14) in practice, we need to choose the tuning parameter λ. Here, we provide a data adaptive way to select λ based on the Bayesian information criterion (BIC) (e.g., Yi et al. (2015)):

$$BIC(\lambda) = -l_j(\lambda) + \log(n)df(\lambda) \qquad (3-18)$$

where $j=1$ or 2, $l_j(\lambda) = \max_{\boldsymbol{\theta}, \boldsymbol{\gamma}} l_j(\boldsymbol{\theta}, \boldsymbol{\gamma})$ is the maximum mixture log-likelihood function for a given tuning parameter λ and $df(\lambda)$ is the model degrees of freedom which is estimated by the sum of the number of nonzero γ values and the number of mixture component parameters (She and Owen, 2011). The optimal tuning parameter λ is chosen by minimizing BIC(λ) over a grid of 100 λ values, equally spaced on the log scale between $\lambda_{\min}$ and $\lambda_{\max}$, where $\lambda_{\max}$ is some large value of λ which corresponds to all zero values of γ_i and $\lambda_{\min}$ is some small value of λ which corresponds to all nonzero values of γ_i.

3.3 Simulation

We conduct several simulation studies to demonstrate the effectiveness of the proposed method and compare it with some of existing estimation methods. We consider two examples: Example 1 is equal variance case and Example 2 is unequal variance case. For both examples, we set the sample size $n=400$. For nonzero γ, the absolute value of γ is generated by a uniform distribution either between 5 and 7 or between 11 and 13. We consider two cases of the proportion of outliers: 5% outliers and 10% outliers by adding nonzero γ_is. The number of replicates is 200 for each simulation setting.

Example 3.1: The samples $(y_1, y_2, \dots, y_n)$ are generated from model (3-3) with $\pi_1=0.3$, $\mu_1=0$, $\pi_2=0.7$, $\mu_2=8$, and $\sigma=1$. The observations $(y_1, y_2, \dots, y_{n1})$ are assigned to the first component (where n_1 is generated by a binomial distribution with $n=400$ and $p=0.3$ and n_1 is the sum of 1's) and the rest of observations, $(y_{n1+1}, \dots, y_n)$, are assigned to the second component. For 5% outliers case (i.e., 20 nonzero γ_is), the first 5 observations are set to be outliers in the first component and the last 15 observations are set to be outliers in the second component; for 10% outliers case (i.e., 40 nonzero γ_is), the first 10 observations are set to be outliers in the first component and the last 30 observations are set to be outliers in the second component.

Example 3.2: The samples $(y_1, y_2, \cdots, y_n)$ are generated from model (3-13) with $\sigma_1 = 1$ and $\sigma_2 = 2$. All other model parameters and simulation settings are the same as in Example 3.1.

3.3.1 Methods and Evaluation Measures

We compare our proposed RMM method using hard and SCAD penalty to one existing robust approach and the traditional MLE. To check the performance of the selection of tuning parameter λ, we also report the "oracle" estimates for both hard and SCAD penalty which are the estimates closest to the true values in the solution path. The seven methods we compared are listed below: Ⅰ. traditional MLE assuming the error has normal density (MLE); Ⅱ. trimmed likelihood estimator (TLE) proposed by Neykov et al. (2007) with the percentage of trimmed data α set to 0.05 ($TLE_{0.05}$); Ⅲ. TLE with the percentage of trimmed data α set to 0.10 ($TLE_{0.10}$); Ⅳ. the proposed RMM using the hard penalty (Hard); Ⅴ. the proposed RMM using the SCAD penalty (SCAD); Ⅵ. the oracle estimate using the hard penalty ($Hard_{oracle}$); Ⅶ. the oracle estimate using the SCAD penalty ($SCAD_{oracle}$).

Note that unlike TLE, the proposed RMM used the data adaptive tuning parameter λ. In addition, unlike our proposed methods, TLE method requires a cutoff value to identify which residuals are outliers. A fixed choice of $\eta = 2.5$ in various situations is applied (Gervini and Yohai, 2002) to identify outliers for TLE method. To evaluate the performance of different estimators, we report the median squared errors (MeSE) of the parameter estimates. Similar to She and Owen (2011), to evaluate the outlier detection performance, we report Ⅰ. the average proportions of masking (M), i.e., the fraction of undetected outliers; Ⅱ. the average proportions of swapping (S), i.e., the fraction of good points labeled as outliers; Ⅲ. the joint detection rate (JD), i.e., the proportion of simulations with 0 masking. Ideally, $M\approx 0$, $S\approx 0$, and $JD\approx 1$.

Note, however, as mentioned in Chapter 2, we must consider label switching issues for mixture models. In our simulation study, the labels are determined by the same way as in Chapter 2.

3.3.2 Results

Simulation results of Example 3.1 are summarized in Table 3.1–Table 3.4. Table 3.1 and Table 3.3 report the three fractions of outlier detection and Table 3.2 and Table 3.4 report the median of squared errors (MeSE) of parameter estimates for each estimation method. For equal variance case, both hard and SCAD have similar results to "oracle" estimators. In case I (5% outliers) with either large $|\gamma|$ or small $|\gamma|$, hard, SCAD, $TLE_{0.05}$, and $TLE_{0.10}$ gain ideal joint outlier detection rate and fraction of undetected true outliers, and small swamping rate but $TLE_{0.10}$ has bigger MeSE of parameter estimates with large $|\gamma|$. In case Ⅱ (10% outliers), hard, SCAD, and $TLE_{0.10}$ get similar performance in terms of both outlier identification and MeSE. $TLE_{0.05}$ fails to work with either large or small $|\gamma|$ due to the smaller α setting (less than the proportion of outliers).

Table 3. 1 ***Outlier Identification Results for Equal Variance Case with Large*** $|\gamma|$

	Hard	Hard_{oracle}	SCAD	SCAD_{oracle}	$\text{TLE}_{0.05}$	$\text{TLE}_{0.10}$
5% outliers						
JD	1. 000	1. 000	1. 000	1. 000	1. 000	1. 000
M	0. 000	0. 000	0. 000	0. 000	0. 000	0. 000
S	0. 000	0. 000	0. 017	0. 023	0. 012	0. 022
10% outliers						
JD	1. 000	1. 000	1. 000	1. 000	0. 010	1. 000
M	0. 000	0. 000	0. 000	0. 000	0. 732	0. 000
S	0. 001	0. 001	0. 042	0. 035	0. 001	0. 013

Table 3. 2 ***MeSE (MSE) of Point Estimates for Equal Variance Case with Large*** $|\gamma|$

	Hard	Hard_{oracle}	SCAD	SCAD_{oracle}	$\text{TLE}_{0.05}$	$\text{TLE}_{0.10}$	MLE
5% outliers							
π	0. 001(0. 001)	0. 001(0. 001)	0. 001(0. 002)	0. 001(0. 003)	0. 001(0. 001)	0. 001(0. 017)	0. 165(0. 220)
μ	0. 007(0. 010)	0. 007(0. 010)	0. 009(0. 017)	0. 009(0. 014)	0. 007(0. 010)	0. 017(3. 159)	38. 05(64. 42)
σ	0. 001(0. 002)	0. 001(0. 002)	0. 002(0. 007)	0. 001(0. 004)	0. 001(0. 001)	0. 024(0. 940)	15. 89(628. 7)
10% outliers							
π	0. 001(0. 001)	0. 001(0. 001)	0. 001(0. 003)	0. 001(0. 003)	0. 840(0. 820)	0. 001(0. 001)	0. 151(0. 236)
μ	0. 008(0. 013)	0. 008(0. 013)	0. 029(0. 039)	0. 040(0. 045)	157. 0(153. 6)	0. 008(0. 013)	40. 61(68. 39)
σ	0. 003(0. 004)	0. 003(0. 004)	0. 012(0. 014)	0. 001(0. 003)	7. 743(7. 729)	0. 001(0. 002)	24. 73(8808)

Table 3. 3 ***Outlier Identification Results for Equal Variance Case with Small*** $|\gamma|$

	Hard	Hard_{oracle}	SCAD	SCAD_{oracle}	$\text{TLE}_{0.05}$	$\text{TLE}_{0.10}$
5% outliers						
JD	0. 990	0. 990	0. 960	1. 000	0. 99	1. 000
M	0. 001	0. 001	0. 030	0. 000	0. 001	0. 000
S	0. 004	0. 004	0. 004	0. 081	0. 013	0. 022

Table 3.3 ***Outlier Identification Results for Equal Variance Case with Small*** $|\gamma|$ **(Continued table)**

	Hard	$Hard_{oracle}$	SCAD	$SCAD_{oracle}$	$TLE_{0.05}$	$TLE_{0.10}$
10% outliers						
JD	0.983	0.985	0.925	0.985	0.165	0.96
M	0.004	0.007	0.050	0.001	0.063	0.004
S	0.033	0.031	0.110	0.112	0.001	0.012

Table 3.4 ***MeSE (MSE) of Point Estimates for Equal Variance Case with Small*** $|\gamma|$

	Hard	$Hard_{oracle}$	SCAD	$SCAD_{oracle}$	$TLE_{0.05}$	$TLE_{0.10}$	MLE
5% outliers							
π	0.003(0.004)	0.003(0.004)	0.001(0.001)	0.003(0.004)	0.001(0.001)	0.002(0.017)	0.003(0.095)
μ	0.023(0.030)	0.035(0.031)	0.156(0.157)	0.041(0.050)	0.009(0.014)	0.022(3.887)	0.251(13.04)
σ	0.018(0.027)	0.002(0.004)	0.404(0.371)	0.016(0.020)	0.001(0.001)	0.022(0.977)	0.539(633.4)
10% outliers							
π	0.001(0.002)	0.001(0.002)	0.001(0.002)	0.001(0.002)	0.001(0.004)	0.001(0.001)	0.003(0.112)
μ	0.017(0.022)	0.028(0.033)	0.026(0.065)	0.026(0.037)	0.143(0.723)	0.010(0.032)	0.806(13.82)
σ	0.019(0.020)	0.004(0.007)	0.034(0.102)	0.031(0.038)	0.261(0.470)	0.001(0.017)	1.315(105.4)

Simulation results of Example 3.2 are summarized in Table 3.5–Table 3.8. Table 3.5 and Table 3.7 report the three fractions of outlier detection and Table 3.6 and Table 3.8 report the median of squared errors (MeSE) of parameter estimates for each estimation method. In case I (5% outliers), Hard, SCAD, $TLE_{0.05}$, and $TLE_{0.10}$ obtain similar outlier identifying rates. Hard, SCAD, and $TLE_{0.05}$ have similar MeSE, while $TLE_{0.10}$ has bigger MeSE for σ. In case Ⅱ (10% outliers), Hard, SCAD, and $TLE_{0.10}$ have the similar outlier identifying rates and MeSE for π and μ but $TLE_{0.10}$ has bigger MeSE for σ with large $|\gamma|$; SCAD fails to work with small $|\gamma|$ but its solution path does include good estimates of the parameters because $SCAD_{oracle}$ has similar results to hard. Therefore, a better method to choose the tuning parameter might be able to improve the performance of SCAD. Like the equal variance case, $TLE_{0.05}$ performs poorly when there are 10% outliers in the data.

Table 3.5 ***Outlier Identification Results for Unequal Variance Case with Large*** $|\gamma|$

	Hard	$Hard_{oracle}$	SCAD	$SCAD_{oracle}$	$TLE_{0.05}$	$TLE_{0.10}$
5% outliers						
JD	1.000	1.000	1.000	1.000	0.967	1.000
M	0.000	0.000	0.000	0.000	0.002	0
S	0.001	0.001	0.003	0.009	0.009	0.031
10% outliers						
JD	1.000	1.000	0.995	0.995	0.000	0.985
M	0.000	0.000	0.005	0.001	0.476	0.000
S	0.001	0.001	0.009	0.013	0.000	0.012

Table 3.6 ***MeSE (MSE) of Point Estimates for Unequal Variance Case with Large*** $|\gamma|$

	Hard	$Hard_{oracle}$	SCAD	$SCAD_{oracle}$	$TLE_{0.05}$	$TLE_{0.10}$	MLE
5% outliers							
π	0.001(0.001)	0.001(0.001)	0.001(0.001)	0.001(0.001)	0.001(0.039)	0.001(0.001)	0.780(0.767)
μ	0.014(0.021)	0.014(0.021)	0.015(0.022)	0.014(0.021)	0.022(17.63)	0.022(0.031)	92.76(91.97)
σ	0.023(0.028)	0.020(0.024)	0.022(0.044)	0.016(0.021)	0.010(0.551)	0.100(0.108)	247.5(243.6)
10% outliers							
π	0.001(0.001)	0.001(0.001)	0.001(0.002)	0.001(0.001)	0.056(0.058)	0.001(0.001)	0.055(0.058)
μ	0.018(0.026)	0.018(0.026)	0.019(0.030)	0.019(0.029)	18.13(18.10)	0.010(0.013)	11.83(11.90)
σ	0.036(0.045)	0.034(0.042)	0.038(0.334)	0.036(0.223)	19.83(19.92)	1.035(1.031)	61.33(61.29)

Table 3.7 ***Outlier Identification Results for Unequal Variance Case with Small*** $|\gamma|$

	Hard	Hard_{oracle}	SCAD	SCAD_{oracle}	$\text{TLE}_{0.05}$	$\text{TLE}_{0.10}$
5% outliers						
JD	0.900	0.900	0.805	1.000	0.875	0.990
M	0.015	0.004	0.182	0	0.011	0.0005
S	0.001	0.001	0.008	0.073	0.008	0.031
10% outliers						
JD	0.800	0.800	0.010	1.000	0.000	0.784
M	0.029	0.027	0.920	0.000	0.268	0.016
S	0.001	0.001	0.000	0.103	0.001	0.008

Table 3.8 ***MeSE (MSE) of Point Estimates for Unequal Variance Case with Small*** $|\gamma|$

	Hard	Hard_{oracle}	SCAD	SCAD_{oracle}	$\text{TLE}_{0.05}$	$\text{TLE}_{0.10}$	MLE
5% outliers							
π	0.001(0.001)	0.001(0.001)	0.001(0.003)	0.001(0.001)	0.001(0.148)	0.001(0.012)	0.192(0.174)
μ	0.017(0.026)	0.017(0.028)	0.045(0.071)	0.022(0.032)	0.025(17.37)	0.024(0.600)	21.97(19.24)
σ	0.009(0.016)	0.004(0.008)	0.183(1.212)	0.004(0.010)	0.013(2.031)	0.100(0.232)	23.76(20.64)
10% outliers							
π	0.001(0.001)	0.001(0.001)	0.027(0.028)	0.001(0.001)	0.161(0.130)	0.001(0.180)	0.248(0.257)
μ	0.021(0.029)	0.022(0.029)	0.086(0.105)	0.025(0.034)	14.45(10.40)	0.008(0.012)	30.41(37.07)
σ	0.016(0.241)	0.008(0.015)	12.04(11.96)	0.010(0.023)	18.54(13.74)	1.017(1.020)	34.52(30.22)

In summary, the proposed Hard has comparable performance to the oracle TLE, that used the correct trimming proportion α, in the simulation studies in terms of both outlier identifying and MeSE (MSE). The proposed SCAD works well for equal variance case. For unequal variance case, SCAD can still work well when there are 5% outliers or the absolute value of γ is big, but does not work properly when the proportion of outliers in data is 10% and the magnitude of

γ is small. A modification on tuning parameter criterion may possibly solve this problem, since the oracle SCAD works well for all cases. The proposed RMM using ℓ_1 norm penalty works with large absolute value of γ when there are 5% outliers but fails to work with small absolute value of γ and more than 5% outliers (The results of soft are omitted here); this agrees with She and Owen (2011). As we expect, the traditional MLE fails to work when there are one or more outliers in the data.

3.4 Real Data Application

We further apply the proposed robust procedure to Acidity dataset (Crawford, 1994; Crawford et al. 1992). The observations are the logarithms of an acidity index measured in a sample 155 lakes in north-central Wisconsin. More details on the data analysis can be found in Crawford (1994), Crawford et al. (1992), and Richardson and Green (1997). Figure 3.1 shows the histogram of Acidity dataset. Based on the result of Richardson and Green (1997), the posterior for three components was largest. Hence we fit this data set by a three-component normal mixture by the traditional MLE and the proposed RMM using HARD penalty.

Table 3.9 reports the parameter estimates on the Acidity data set. For the original data where there are no outliers, the proposed Hard has similar parameter estimates to that of the traditional MLE. To see the effects of outliers on Hard and MLE, similar to McLachlan and Peel (2000), we add one outlier ($y=12$) to the

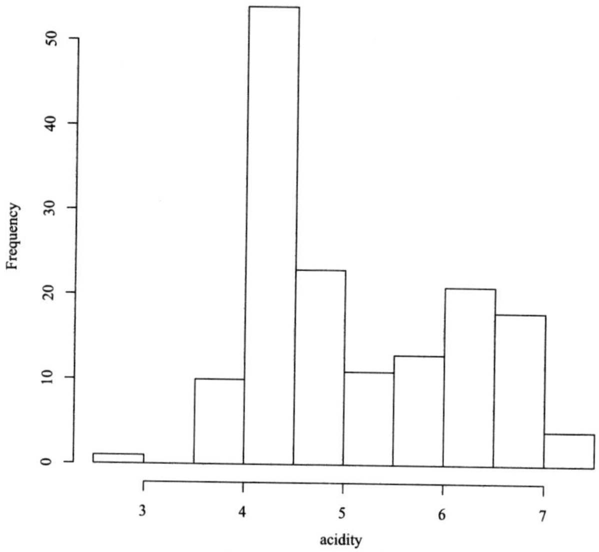

Figure 3. 1 ***Histogram for Acidity data***

original data. Based on Table 3. 9, the proposed Hard is not influenced by the outlier and gives similar parameter estimates to the case of no outliers. However, MLE gives different parameter estimates from the case of no outliers. In addition, note that MLE provides the same component means for the first and second components. We further add three identical outliers ($y = 12$) to the data. As we expect, Hard still provides similar estimates to the case of no outliers. However, MLE fits a new component to the outliers and gives totally different estimates from the case of no outliers.

Table 3.9 ***Parameter Estimation on Acidity Data Set***

		π_1	π_2	π_3	μ_1	μ_2	μ_3	σ
MLE	No outliers	0.589	0.138	0.273	4.320	5.682	6.504	0.365
	1 outlier	0.327	0.324	0.349	4.455	4.455	6.448	0.687
	3 outliers	0.503	0.478	0.019	5.105	5.105	12.00	1.028
Hard	No outliers	0.588	0.157	0.255	4.333	5.720	6.545	0.336
	1 outlier	0.591	0.157	0.252	4.333	5.723	6.548	0.334
	3 outliers	0.597	0.157	0.246	4.333	5.729	6.553	0.331

3.5 Discussion

The main contribution of this paper is to propose a robust mixture via mean shift penalization model (RMM). In addition, we proposed a thresholding embedded EM algorithm to find the proposed robust estimate. Based on the simulation studies and real data analysis, we can see that RMM with Hard penalty has similar performance to TLE that uses an oracle trimming proportion. Note, however, RMM can adaptively choose the tuning parameter λ based on BIC. In addition, the proposed RMM can naturally detect outliers corresponding to nonzero γ_is.

In this chapter, we mainly focus on normal mixture model. We think the proposed robust procedure RMM can be also extended to other mixture models, such as mixtures of binomial and mixtures of poisson. In addition, the proposed RMM can be also extended to mixture of linear regression models and mixture of generalized linear models.

Chapter 4
Outlier Detection and Robust Mixture Regression Using Nonconvex Penalized Likelihood

4.1 Introduction

As mentioned in Chapter 2, although the finite mixture models with the maximum likelihood inference have greatly enriched the toolkit of regression analysis, the model is very sensitive to outliers. In this chapter, we propose a Robust Mixture Regression via Mean shift penalization approach (RMRM or RM^2), to conduct simultaneous outlier detection/accomodation and robust parameter estimation in finite normal mixture regression models. Our method is motivated by She and Owen (2011) and Lee et al. (2012), in which penalized estimation methods were adopted to induce the sparsity of a case-specific parameter vector for accommodating outliers in linear regression models. This method generalizes the robust mixture model proposed in Chapter 3 and can handle more general supervised learning tasks. Under the general framework of mixture regression, there are several new challenges for adopting the nonconvex penalization methods. For example, the problem of maximizing the likelihood itself becomes a nonconvex problem, which complicates the computation. When the components have unequal variances, the simple mean shift model will not work well since the definition of an outlier may become ambiguous as the scale of the outlying effect of a partic-ular point may vary across different components. We propose to add a component specific mean-shift term for each component and for each observation and these terms are designed to be proportional to the component variances, accounting for the potential heteroscedasticity among different components. We propose an efficient iterative thresholding embeded EM algorithm to solve the nonconvex RM^2 problem, and our proposed estimator is demonstrated to be highly robust against gross outliers and leverage points.

The rest of chapter 4 is organized as follows. In Section 4. 2, we propose the RM^2 approach. In Section 4. 3, we compare the proposed methods to several existing methods via simulation studies. A real application showcasing the efficacy of the proposed method is presented in Section 4. 4, and we conclude the chapter in Section 4. 5.

4.2 Robust Mixture Regression via Mean-shift Penalization

To illustrate the main idea, we start from the simple case that the mixture components have equal variances, i.e., $\sigma_1^2=\cdots=\sigma_m^2=\sigma^2$. Motivated by the mean-shift linear regression model considered by She and Owen (2011) and Lee et al. (2012), it is natural to consider the following mixture model with a mean-shift parameterization, i.e.,

$$f(y_i;\boldsymbol{\theta},\gamma_i)=\sum_{j=1}^{m}\pi_j\phi(y_i;\boldsymbol{x}_i^T\boldsymbol{\beta}_j+\gamma_i,\sigma^2),\qquad i=1,\ldots,n \quad (4\text{-}1)$$

where $\boldsymbol{\theta}=(\pi_1,\boldsymbol{\beta}_1,\ldots,\pi_m,\boldsymbol{\beta}_m,\sigma)^T$. Here, for each observation, a shift parameter, γ_i, is added to its mixture mean structure; we thus refer to the above as a mean shifted mixture model. Without any further constraints on the model parameters, it is obvious that the mean-shift model is over-parameterized and hence the parameters are not fully identifiable. The essence of this formulation lies in the sparsity assumption on γ_i, i.e., we shall assume many γ_is are in fact zero, corresponding to the normal observations, and only a few γ_is are nonzero, corresponding to the outlying observations. Therefore, promoting sparsity of γ_i in model estimation provides a direct way for identifying and accommodating outliers in the mixture model.

Now consider the general case that the mixture components are allowed to have unequal variances, i. e., $\epsilon_j\sim N(0,\sigma_j^2)$. This heteroscedasticity of component variances imposes additional challenges for identifying outliers, as the definition of an "outlier" even becomes ambiguous due to the fact that the mixture components are of different scales. In the general setting, whether an observation is an outlier to a certain component should be judged based on the scale of that component, and an observation should be declared as an outlier only if it is far away from all the centroids of the mixture components. This motivates us to further extend model (4-1) to take into account the scaling issue. The main idea is to make the case-specific mean shift parameter γ_i be scale invariant, so that the magnitude of γ_i itself represents the standardized distance from the observation to all the cluster centroids. We thus propose the following robust mixture regression model with mean-shift (RM^2),

$$f(y_i \mid \boldsymbol{x}_i, \boldsymbol{\theta}, \gamma_i) = \sum_{j=1}^{m} \pi_j \phi(y_i; \boldsymbol{x}_i^T \boldsymbol{\beta}_j + \gamma_i \sigma_j, \sigma_j^2), \qquad i = 1, \ldots, n \tag{4-2}$$

where we redefine $\boldsymbol{\theta}=(\pi_1, \beta_1, \sigma_1, \ldots, \pi_m, \beta_m, \sigma_m)^T$. The outlying effect is made both case-specific and component-specific, i.e., the outlying effect of the ith observation to the jth component is modeled by $\gamma_i\sigma_j$, depending directly on the scale of the jth component. In this way, γ_i becomes scale free, and can be simply understood as the number of standard deviations shifted from the correct component mean structures. Similar to that of the traditional mixture model (2.1), the above penalized log-likelihood is also unbounded. That is, the penalized log-likelihood goes to infinity when $y_i = \boldsymbol{x}_i^T \beta_j + \gamma_{ij}\sigma_j$, and $\sigma_j \to 0$ (Hathaway, 1985, 1986; Chen et al. 2008; Yao, 2010). To circumvent this problem, as stated in chapter 2, we put some constraints on the parameter space such that the component variance has some low limit following Hathaway (1985, 1986).

The efficient and accurate recovery of the sparse vector $\boldsymbol{\gamma}=(\gamma_1, \ldots, \gamma_n)^T$ holds the key to realize the bearing of the powerful framework of the proposed mean shifted mixture model. In recent years, the penalized estimation approach has undergone exciting developments for sparse learning and variable selection. This motivates us to consider a penalized likelihood approach. Given a random sample $\{(\boldsymbol{x}_i, y_i), i=1, 2, \ldots, n\}$ from model (4-2), the log-likelihood function is given by

$$\ell_n(\boldsymbol{\theta},\boldsymbol{\gamma})=\sum_{i=1}^{n}\log\left\{\sum_{j=1}^{m}\pi_j\phi(y_i-\gamma_i\sigma_j-\boldsymbol{x}_i^T\boldsymbol{\beta}_j;0,\sigma_j^2)\right\}$$

We propose a penalized likelihood approach to conduct model estimation and outlier detection,

$$pl_n(\boldsymbol{\theta},\boldsymbol{\gamma})=\ell_n(\boldsymbol{\theta},\boldsymbol{\gamma})-\sum_{i=1}^{n}P_\lambda(|\gamma_i|) \qquad (4-3)$$

where $P_\lambda(\cdot)$ is some penalty function chosen to induce the sparsity in $\boldsymbol{\gamma}$, with λ being a tuning parameter controlling the degrees of penalization (She and Owen, 2011). There are many choices for the penalty function in the above criterion. We mainly focus on the ℓ_1, ℓ_0 and SCAD penalties which have respectively same forms as described in chapter 3. Each of these penalty forms corresponds to certain thresholding rule, thus capable of performing shrinkage and producing exact zero solution, e.g., ℓ_1 penalty corresponds to a soft-thresholding rule, ℓ_0 penalty corresponds to a hard-thresholding rule, and SCAD penalty corresponds to a SCAD-thresholding rule.

In classical mixture regression problem, the EM algorithm is commonly used to maximize the likelihood, as the component labels are unobservable and can be treated as missing data. Here, we propose an iterative thresholding embeded EM algorithm to maximize the proposed penalized log-likelihood criterion. Define the latent variable z_{ij} and denote the complete data by $\{(\boldsymbol{x}_i, z_i, y_i), i=1, 2, \ldots, n\}$, where the component labels $\boldsymbol{z}_i=(z_{i1}, z_{i2}, \cdots, z_{im})$ are not observable in practice. The penalized complete log-likelihood function is

$$pl_n^c(\boldsymbol{\theta}, \boldsymbol{\gamma}) = \ell_n^c(\boldsymbol{\theta}, \boldsymbol{\gamma}) - \sum_{i=1}^{n} P_\lambda(|\gamma_i|) \tag{4-4}$$

where the complete log-likelihood is given by

$$\ell_n^c(\boldsymbol{\theta}, \boldsymbol{\gamma}) = \sum_{i=1}^{n}\sum_{j=1}^{m} z_{ij} \log\left\{\pi_j \phi(y_i - \gamma_i\sigma_j - \boldsymbol{x}_i^T\boldsymbol{\beta}_j; 0, \sigma_j^2)\right\} \tag{4-5}$$

In the E-step, the conditional expectation of the penalized complete log-likelihood (4-4) is computed, and we then maximize it with respect to $\boldsymbol{\theta}$ and $\boldsymbol{\gamma}$ in the M-step. Specifically, in the M-step, we alternatingly update $\boldsymbol{\theta}$ and $\boldsymbol{\gamma}$ with the other part held fixed, until convergence is reached. For fixed $\boldsymbol{\gamma}$, both p_{ij} and $\boldsymbol{\beta}$ can be solved explicitly. As each σ_j appears in the mean structure, it no longer has an explicit solution. However, the estimation of each σ_j is separable so that the problem is easily solvablc by standard optimization method such as Newton-Raphson. For fixed p_{ij}, $\boldsymbol{\beta}$, and σ_j, $\boldsymbol{\gamma}$ is updated by maximizing

$$\sum_{i=1}^{n}\sum_{j=1}^{m} p_{ij}^{(k+1)} \log \phi(y_i - \gamma_i\sigma_j - \boldsymbol{x}_i^T\boldsymbol{\beta}_j; 0, \sigma_j^2) - \sum_{i=1}^{n} P_\lambda(|\gamma_i|)$$

It can be shown that the above problem is separable in each γ_i, for which it suffices to minimize

$$\frac{1}{2}\left[\left\{\gamma_i - \sum_{j=1}^{m} \frac{p_{ij}^{(k+1)}}{\sigma_j}(y_i - \boldsymbol{x}_i^T\boldsymbol{\beta}_j)\right\}^2\right] + P_\lambda(|\gamma_i|) \tag{4-6}$$

The thresholding rules for soft, hard, and SCAD are given, respectively, as follows:

$$\gamma_i = \Theta_{soft}(\xi_i; \lambda) = \begin{cases} 0, & \text{if } |\xi_i| \leqslant \lambda \\ \xi_i - sgn(\xi_i)\lambda, & \text{if } |\xi_i| > \lambda \end{cases}$$

$$\gamma_i = \Theta_{hard}(\xi_i;\lambda) = \begin{cases} 0, & \text{if } |\xi_i| \leqslant \lambda \\ \xi_i, & \text{if } |\xi_i| > \lambda \end{cases}$$

and

$$\gamma_i = \Theta_{SCAD}(\xi_i;\lambda) = \begin{cases} sgn(\xi_i)(|\xi_i| - \lambda)_+, & \text{if } |\xi_i| \leqslant 2\lambda \\ \frac{(a-1)\xi_i - sgn(\xi_i)a\lambda}{a-2}, & \text{if } 2\lambda < |\xi_i| \leqslant a\lambda \\ \xi_i, & \text{if } |\xi_i| > a\lambda \end{cases}$$

where

$$\xi_i = \sum_{j=1}^{m} \frac{p_{ij}}{\sigma_j}(y_i - \mathbf{x}_i^T \boldsymbol{\beta}_j)$$

The detailed proposed thresholding embeded EM algorithm to maximize the penalized log-likelihood (4-3) is summarized in Algorithm 4. 1. Based on the property of EM algorithm, for any fixed tuning parameter $\boldsymbol{\lambda}$, each iteration of the E-step and M-step of Algorithm 4. 1 monotonically non-decreases the penalized log-likelihood function, i.e., $pl_n(\hat{\boldsymbol{\theta}}^{(k+1)}, \hat{\boldsymbol{\gamma}}^{(k+1)}) \geqslant pl_n(\hat{\boldsymbol{\theta}}^{(k)}, \hat{\boldsymbol{\gamma}}^{(k)})$, *for all* $k \geqslant 0$.

Algorithm 4. 1: *Initialize* $\boldsymbol{\theta}^{(0)}$ *and* $\boldsymbol{\gamma}^{(0)}$. *Set* $k \leftarrow 0$.

E-Step: *Compute the conditional expectation*:

$$Q(\boldsymbol{\theta}, \boldsymbol{\gamma} \mid \boldsymbol{\theta}^{(k)}, \boldsymbol{\gamma}^{(k)}) = \sum_{i=1}^{n}\sum_{j=1}^{m} p_{ij}^{(k+1)} \left[\log \pi_j + \log \phi(y_i - \gamma_i\sigma_j - \boldsymbol{x}_i^T \boldsymbol{\beta}_j; 0, \sigma_j^2)\right] - \sum_{i=1}^{n} P_\lambda(|\gamma_i|)$$

where

$$p_{ij}^{(k+1)} = E(z_{ij}|y_i;\boldsymbol{\theta}^{(k)}) = \frac{\pi_j^{(k)}\phi(y_i - \gamma_i^{(k)}\sigma_j^{(k)} - \boldsymbol{x}_i^T\boldsymbol{\beta}_j^{(k)};0,\sigma_j^{2(k)})}{\sum_{j=1}^m \pi_j^{(k)}\phi(y_i - \gamma_i^{(k)}\sigma_j^{(k)} - \boldsymbol{x}_i^T\boldsymbol{\beta}_j^{(k)};0,\sigma_j^{2(k)})}$$

M-Step: *Update* $\pi_j^{(k+1)} = \frac{\sum_{i=1}^n p_{ij}^{(k+1)}}{n}$ *and update other parameters by maximizing* $Q(\boldsymbol{\theta}, \boldsymbol{\gamma}|\boldsymbol{\theta}^{(k)}, \boldsymbol{\gamma}^{(k)})$, *i.e., start from* $(\boldsymbol{\beta}^{(k)}, \sigma_j^{2(k)}, \boldsymbol{\gamma}^{(k)})$ *and iterate the following steps until convergence to obtain* $(\boldsymbol{\beta}^{(k+1)}, \sigma_j^{2(k+1)}, \boldsymbol{\gamma}^{(k+1)})$:

$$\boldsymbol{\beta}_j \leftarrow \left(\sum_{i=1}^n \boldsymbol{x}_i\boldsymbol{x}_i^T p_{ij}^{(k+1)}\right)^{-1}\left(\sum_{i=1}^n \boldsymbol{x}_i p_{ij}^{(k+1)}(y_i - \gamma_i\sigma_j)\right)$$

$$\sigma_j^2 \leftarrow \arg\max_{\sigma_j^2}\sum_{i=1}^n p_{ij}^{(k+1)}\log\phi(y_i - \gamma_i\sigma_j - \boldsymbol{x}_i^T\boldsymbol{\beta}_j;0,\sigma_j^2)$$

$$\gamma_i \leftarrow \Theta(\xi_i;\lambda)$$

where Θ *denotes a thresholding rule depending on the penalty form adopted, and* $\xi_i = \sum_{j=1}^m \frac{p_{ij}^{(k+1)}}{\sigma_j}(y_i - \boldsymbol{x}_i^T\beta_j)$.

The proposed EM algorithm is for any fixed tuning parameter λ. In practice, we need to choose an optimal λ and hence an optimal set of parameter estimates. We construct a Bayesian information criterion (BIC) for tuning parameter selection (e.g., Yi et al. (2015)),

$$BIC(\lambda) = -\ell(\lambda) + \log(n)df(\lambda) \tag{4-7}$$

where $\ell(\lambda)$ is the mixture log-likelihood function evaluated at the parameter estimates with tuning parameter λ and $df(\lambda)$ is the degrees of freedom of the resulting model. Following Zou (2006), we estimate the degrees of freedom using the sum of the number of nonzero elements of the vector $\hat{\boldsymbol{\gamma}}(\lambda)$ and the number of component parameters in the mixture model. We fit the model for 100 λ values equally spaced at the log scale in an interval $(\lambda_{\min}, \lambda_{\max})$, where $\lambda_{\min}$ is some λ value for which about 50% of the entries in γ are nonzero, and $\lambda_{\max}$ corresponds to some λ value for which $\boldsymbol{\gamma}$ is estimated as a zero vector.

We note that from outlier detection point of view or for practical consideration, there may be other methods to determine the λ value or choose the optimal solution along the solution path. For example, based on prior knowledge, one may decide to discard 5% of the observations as outliers; then a solution with approximately 5% of nonzero $\boldsymbol{\gamma}$ values can be chosen as the final solution. In the scale-invariate model, as $\boldsymbol{\gamma}$ can be interpreted as the number of standard deviations from the mean structure, one may also examine the magnitude of the $\boldsymbol{\gamma}$ estimates to determine the number of possible outliers. Although we mainly use BIC in this paper, we shall see that by formulating the outlier detection problem as a penalized regression method, the many well-studied model selection criteria including C_p, AIC, and GCV are all applicable.

4.3 Simulation

4.3.1 Simulation Setups

We consider two mixture model setups, in which the observations are contaminated with additive outliers, to evaluate the final sample performance of the proposed approach and compare it with several existing methods.

Example 4.1: For each $i=1, \ldots, n$, y_i is independently generated by

$$y_i=\begin{cases}1-x_{1i}+x_{2i}+\gamma_i\sigma+\epsilon_{i1}, \text{if } z_{i1}=1\\ 1+3x_{1i}+x_{2i}+\gamma_i\sigma+\epsilon_{i2}, \text{ if } z_{i1}=0\end{cases}$$

where z_{i1} is a component indicator generated from Bernoulli distribution with $P(z_{i1}=1)=0.3$; x_{1i} and x_{2i} are independently generated from $N(0, 1)$, and the error terms ϵ_{i1} and ϵ_{i2} are also independently generated from $N(0, \sigma^2)$ with $\sigma^2=1$.

Example 4.2: For each $i=1, \ldots, n$, y_i is independently generated by

$$y_i=\begin{cases}1-x_{1i}+x_{2i}+\gamma_i\sigma_1+\epsilon_{i1}, \text{if } z_{i1}=1\\ 1+3x_{1i}+x_{2i}+\gamma_i\sigma_2+\epsilon_{i2}, \text{ if } z_{i1}=0\end{cases}$$

where z_{i1} is a component indicator generated from Bernoulli distribution with $P(z_{i1}=1)=0.3$; x_{1i} and x_{2i} are independently generated from $N(0, 1)$, and the error terms ϵ_{i1} and ϵ_{i2} are independently generated from $N(0, \sigma_1^2)$ and $N(0, \sigma_2^2)$, respectively, with $\sigma_1^2=1$ and $\sigma_2^2=4$.

We consider two magnitudes of outliers, i.e., the absolute value of any non-zero mean shift parameter, $\alpha_i = |\gamma_i|$, is generated from uniform distribution either between 5 and 7 or between 11 and 13. We consider two proportions of outliers, either 5% or 10%. In each setting, the sample size is set to be $n=400$ and we repeat the simulation 200 times. Specifically, in Example 4.1, we first generate $n=400$ observations according to Model 1 with all γ_is set to be zero; when there are 5% (10%) outliers, 5 (10) observations from the first component are then replaced by $y_i=1-x_{1i}+x_{2i}+\gamma_i+\epsilon_{1i}$ with $x_{1i}=2$, $x_{2i}=2$, and $\gamma_i=-\alpha_i$, and 15

(30) observations from the second component are replaced by $y_i = 1+3x_{1i}+ x_{2i}+ \gamma_i+\epsilon_{2i}$ with $x_{1i}=2$, $x_{2i}=2$, and $\gamma_i=\alpha_i$. In Example 4.2, the additive outliers are generated in exactly the same fashion as in Example 4.1, and the only difference is that the component variances are unequal in the latter example.

4.3.2 Methods and Evaluation Measures

We compare our proposed RM^2 approach with soft, hard, and SCAD penalties to three existing robust approaches and the traditional normal mixture model. To alleviate the inaccuracy in tuning parameter selection and examine the true potential of the proposed approaches, we also report the "oracle" penalized regression estimator for each penalty, which is defined as the solution whose number of selected outliers is the smallest number greater than or equal to the number of true outliers on the solution path. These are the estimators we would have obtained if the true proportion of outliers is known a priori. The eleven methods we compared are listed below: Ⅰ. the traditional MLE in mixture linear regression with normally distributed error (MLE); Ⅱ. trimmed likelihood estimator (TLE) proposed by Neykov et al. (2007) with the percentage of trimmed data α set to 0.05 ($TLE_{0.05}$); Ⅲ. TLE with the percentage of trimmed data α set to 0.10 ($TLE_{0.10}$); Ⅳ. the robust estimator based on an modified EM algorithm with bisquare loss (MEM-bisquare) proposed by Bai et al. (2012); Ⅴ. the MLE in mixture linear regression assuming t-distributed error (Mixregt); Ⅵ. the proposed RM^2 using the hard penalty (Hard); Ⅶ. the proposed RM^2 using the SCAD penalty (SCAD); Ⅷ. the proposed RM^2 using the soft penalty (Soft); Ⅸ. the oracle estimate using the hard penalty ($Hard_{oracle}$); Ⅹ. the oracle estimate using the SCAD penalty ($SCAD_{oracle}$) ; XI. the oracle estimate using the soft penalty ($Soft_{oracle}$).

As mentioned in Chapter 2, we must consider label switching issues for mixture models. In our simulation study, the labels are determined by the same way as described in Chapter 2. To evaluate the estimation performance, we report both the median squared errors (MeSE) and the mean squared errors (MSE) of the parameter estimates. Similar to Chapter 3, to evaluate the outlier detection performance, we report three measures: the average proportion of masking (M), i.e., the fraction of undetected outliers, the average proportion of swapping (S), i.e., the fraction of good points labeled as outliers, and the joint detection rate (JD), i.e., the proportion of simulations with 0 masking. The simulation results are summarized in Table 4. 1–Table 4. 8.

4. 3. 3 Results

The simulation results of Example 4. 1 (equal variance case) are reported in Table 4. 1 – Table 4. 4. Table 4. 1 and Table 4. 3 report the three fractions of outlier detection and Table 4. 2 and Table 4. 4 report the median of squared errors (MeSE) of parameter estimates for each estimation method. In the case of 5% outliers, all methods gain ideal outlier detection rates and small MeSE of parameter estimates with large $|\gamma|$; all methods except for Soft have high joint outlier detection rate and small MeSE with small $|\gamma|$. In the case of 10% outliers, Hard, SCAD, and $TLE_{0.10}$ work well in terms of both outlier detection rates and MeSE with large $|\gamma|$, whereas Soft, $TLE_{0.05}$, MEM–bisquare, and Mixregt have low joint outlier detection rates and big MeSE; with small $|\gamma|$, $TLE_{0.10}$, and

Mixregt work better than other methods in outlier identification but Hard, Hard_{oracle}, and SCAD_{oracle} obtain similar MeSE to those of $\text{TLE}_{0.10}$ and Mixregt.

Table 4.1 ***Outlier Identification Results for Equal Variance Case with Large*** $|\gamma|$

	5% outliers			10% outliers		
	M	S	JD	M	S	JD
Hard	0.000	0.001	1.000	0.000	0.002	1.000
Hard_{oracle}	0.000	0.001	1.000	0.000	0.000	1.000
SCAD	0.005	0.014	0.995	0.001	0.003	0.994
SCAD_{oracle}	0.000	0.031	1.000	0.000	0.002	1.000
Soft	0.066	0.017	0.920	0.840	0.005	0.000
Soft_{oracle}	0.000	0.033	1.000	0.179	0.024	0.375
$\text{TLE}_{0.05}$	0.000	0.007	1.000	0.749	0.050	0.000
$\text{TLE}_{0.10}$	0.000	0.003	1.000	0.000	0.007	1.000
MEM-bisquare	0.000	0.005	1.000	0.639	0.061	0.145
Mixregt	0.000	0.078	1.000	0.313	0.096	0.555

Table 4.2 ***MeSE (MSE) of Point Estimates for Equal Variance Case with Large*** $|\gamma|$

	π		β		σ	
5% outliers						
	MSE	MeSE	MSE	MeSE	MSE	MeSE
Hard	0.002	0.001	0.058	0.042	0.020	0.005
Hard_{oracle}	0.002	0.001	0.050	0.024	0.024	0.004
SCAD	0.003	0.001	0.053	0.042	0.018	0.005
SCAD_{oracle}	0.002	0.001	0.049	0.041	0.004	0.002
Soft	0.010	0.006	0.771	0.193	1.957	0.045
Soft_{oracle}	0.007	0.005	0.126	0.119	0.462	0.459

Table 4.2 ***MeSE (MSE) of Point Estimates for Equal Variance Case with Large*** $|\gamma|$ **(Continued table)**

	π		β		σ	
5% outliers						
	MSE	MeSE	MSE	MeSE	MSE	MeSE
$TLE_{0.05}$	0.002	0.001	0.047	0.037	0.002	0.001
$TLE_{0.10}$	0.002	0.001	0.085	0.067	0.025	0.023
MEM-bisquare	0.002	0.001	0.050	0.041	0.007	0.004
Mixregt	0.003	0.002	0.090	0.080	0.123	0.121
MLE	0.470	0.680	17.20	20.33	2.912	2.920
10% outliers						
	MSE	MeSE	MSE	MeSE	MSE	MeSE
Hard	0.002	0.001	0.059	0.047	0.015	0.006
$Hard_{oracle}$	0.001	0.001	0.071	0.044	0.002	0.002
SCAD	0.002	0.001	0.088	0.047	0.037	0.005
$SCAD_{oracle}$	0.001	0.001	0.012	0.045	0.002	0.002
Soft	0.293	0.409	17.85	19.64	3.308	3.026
$Soft_{oracle}$	0.065	0.017	11.96	6.176	2.195	2.518
$TLE_{0.05}$	0.274	0.046	50.94	49.08	0.298	0.275
$TLE_{0.10}$	0.002	0.001	0.057	0.046	0.002	0.001
MEM-bisquare	0.279	0.043	39.81	45.74	0.143	0.120
Mixregt	0.212	0.005	18.05	0.174	0.058	0.056
MLE	0.075	0.014	11.55	10.09	4.462	4.459

Table 4.3 ***Outlier Identification Results for Equal Variance Case with Small*** $|\gamma|$

	5% outliers			10% outliers		
	M	S	JD	M	S	JD
Hard	0.002	0.001	0.965	0.038	0.001	0.615
$Hard_{oracle}$	0.000	0.060	1.000	0.005	0.001	0.790

Table 4.3 ***Outlier Identification Results for Equal Variance Case with Small*** $|\gamma|$ **(Continued table)**

	5% outliers			10% outliers		
	M	S	JD	M	S	JD
SCAD	0.001	0.001	0.950	0.957	0.001	0.000
$SCAD_{oracle}$	0.001	0.054	0.985	0.119	0.059	0.575
Soft	0.906	0.000	0.000	0.959	0.001	0.000
$Soft_{oracle}$	0.002	0.054	0.955	0.263	0.031	0.000
$TLE_{0.05}$	0.002	0.008	0.965	0.490	0.015	0.000
$TLE_{0.10}$	0.000	0.026	0.995	0.002	0.007	0.945
MEM-bisquare	0.038	0.008	0.865	0.471	0.019	0.200
Mixregt	0.000	0.074	0.990	0.007	0.050	0.930

Table 4.4 ***MeSE (MSE) of Point Estimates for Equal Variance Case with Small*** $|\gamma|$

	π		β		σ	
	5% outliers					
	MSE	MeSE	MSE	MeSE	MSE	MeSE
Hard	0.002	0.001	0.067	0.048	0.013	0.006
$Hard_{oracle}$	0.002	0.001	0.057	0.048	0.005	0.002
SCAD	0.003	0.001	0.116	0.087	0.016	0.008
$SCAD_{oracle}$	0.002	0.001	0.080	0.068	0.006	0.003
Soft	0.003	0.001	1.056	1.031	0.281	0.259
$Soft_{oracle}$	0.002	0.001	0.306	0.286	0.058	0.054
$TLE_{0.05}$	0.002	0.001	0.060	0.054	0.002	0.001
$TLE_{0.10}$	0.002	0.001	0.093	0.086	0.027	0.026
MEM-bisquare	0.003	0.001	1.237	0.058	0.009	0.004
Mixregt	0.002	0.001	0.102	0.089	0.120	0.121

Table 4.4 ***MeSE (MSE) of Point Estimates for Equal Variance Case with Small*** $|\gamma|$ **(Continued table)**

	π		β		σ	
	10% outliers					
	MSE	MeSE	MSE	MeSE	MSE	MeSE
MLE	0.003	0.001	1.091	1.078	0.315	0.308
Hard	0.002	0.001	0.134	0.046	0.015	0.006
Hard_{oracle}	0.001	0.001	0.057	0.049	0.008	0.007
SCAD	0.002	0.001	2.310	2.273	0.563	0.565
SCAD_{oracle}	0.002	0.001	0.784	0.129	0.131	0.008
Soft	0.003	0.001	2.357	2.322	0.538	0.539
Soft_{oracle}	0.003	0.001	1.734	1.685	0.344	0.340
$\text{TLE}_{0.05}$	0.015	0.003	7.472	0.937	0.142	0.141
$\text{TLE}_{0.10}$	0.002	0.001	0.058	0.050	0.002	0.001
MEM-bisquare	0.029	0.004	7.397	1.314	0.134	0.103
Mixregt	0.005	0.001	0.247	0.111	0.074	0.074
MLE	0.003	0.001	2.386	2.347	0.576	0.567

Table 4.5-Table 4.8 summarize the simulation results of Example 4.2 (unequal variance case). Table 4.5 and Table 4.7 show the three fractions of outlier detection and Table 4.6 and Table 4.8 show the median of squared errors (MeSE) of parameter estimates for each estimation method. All methods except for soft have high joint outlier detection rates when the proportion of outliers is 5% with large $|\gamma|$; SCAD and MEM-bisquare have low joint detection rates with small γ but SCAD_{oracle} has similar performance to Hard. When there are 10% out-

liers in the data, Hard, SCAD, and $TLE_{0.10}$ have outstanding performance in terms of outlier identification rates and MeSE with large $|\gamma|$; Hard and SCAD do not work well in terms of joint outlier detection rate with small $|\gamma|$ but MeSE of $Hard_{oracle}$ is comparable to $TLE_{0.10}$. Mixregt has low joint detection rate with large $|\gamma|$ and high JD rate with small $|\gamma|$ for 10% outliers case.

Table 4.5 ***Outlier Identification Results for Unequal Variance Case with Large*** $|\gamma|$

	5% outliers			10% outliers		
	M	S	JD	M	S	JD
Hard	0.000	0.001	1.000	0.000	0.001	1.000
$Hard_{oracle}$	0.000	0.001	1.000	0.000	0.000	1.000
SCAD	0.002	0.004	0.995	0.003	0.005	0.990
$SCAD_{oracle}$	0.000	0.010	1.000	0.000	0.005	1.000
Soft	0.894	0.005	0.050	0.960	0.001	0.000
$Soft_{oracle}$	0.005	0.225	0.995	0.728	0.084	0.010
$TLE_{0.05}$	0.004	0.008	0.915	0.656	0.018	0.000
$TLE_{0.10}$	0.008	0.032	0.845	0.003	0.008	0.900
MEM-bisquare	0.062	0.006	0.915	0.722	0.012	0.010
Mixregt	0.000	0.078	1.000	0.461	0.097	0.200

Table 4.6 ***MeSE (MSE) of Point Estimates for Unequal Variance Case with Large*** $|\gamma|$

	π		β		σ	
	5% outliers					
	MSE	MeSE	MSE	MeSE	MSE	MeSE
Hard	0.003	0.001	0.104	0.091	0.036	0.028
$Hard_{oracle}$	0.003	0.001	0.102	0.091	0.029	0.025
SCAD	0.004	0.001	0.129	0.096	0.114	0.025
$SCAD_{oracle}$	0.003	0.001	0.112	0.095	0.029	0.016

Table 4.6 ***MeSE (MSE) of Point Estimates for Unequal Variance Case with Large*** **$|\gamma|$ (Continued table)**

	π		β		σ	
	5% outliers					
	MSE	MeSE	MSE	MeSE	MSE	MeSE
Soft	0. 689	0. 757	36. 76	37. 74	160. 0	182. 2
Soft_{oracle}	0. 011	0. 004	0. 405	0. 196	1. 574	0. 556
$\text{TLE}_{0.05}$	0. 077	0. 002	9. 160	0. 096	0. 502	0. 023
$\text{TLE}_{0.10}$	0. 259	0. 007	1. 528	0. 219	1. 756	0. 655
MEM-bisquare	0. 087	0. 004	9. 835	0. 115	0. 637	0. 102
Mixregt	0. 008	0. 003	0. 421	0. 182	0. 683	0. 655
MLE	0. 761	0. 763	43. 20	41. 84	186. 2	186. 5
	10% outliers					
	MSE	MeSE	MSE	MeSE	MSE	MeSE
Hard	0. 003	0. 001	0. 122	0. 100	0. 060	0. 052
Hard_{oracle}	0. 003	0. 001	0. 122	0. 100	0. 056	0. 044
SCAD	0. 006	0. 002	0. 319	0. 115	1. 837	0. 044
SCAD_{oracle}	0. 003	0. 002	0. 205	0. 108	0. 094	0. 046
Soft	0. 587	0. 590	39. 49	38. 87	193. 5	194. 2
Soft_{oracle}	0. 570	0. 589	46. 68	45. 69	110. 7	112. 9
$\text{TLE}_{0.05}$	0. 654	0. 679	98. 20	90. 68	1. 960	1. 970
$\text{TLE}_{0.10}$	0. 063	0. 002	10. 37	0. 125	0. 403	0. 018
MEM-bisquare	0. 622	0. 652	94. 93	86. 53	2. 397	2. 291
Mixregt	0. 516	0. 638	70. 46	81. 49	0. 968	0. 998
MLE	0. 593	0. 593	40. 89	38. 98	188. 1	195. 2

Table 4.7 ***Outlier Identification Results for Unequal Variance Case with Small*** $|\gamma|$

	5% outliers			10% outliers		
	M	S	JD	M	S	JD
Hard	0.003	0.001	0.955	0.649	0.000	0.125
$Hard_{oracle}$	0.001	0.001	0.995	0.051	0.006	0.725
SCAD	0.828	0.001	0.055	0.951	0.001	0.000
$SCAD_{oracle}$	0.001	0.070	0.980	0.300	0.061	0.215
Soft	0.889	0.001	0.000	0.952	0.001	0.000
$Soft_{oracle}$	0.000	0.233	1.000	0.423	0.050	0.000
$TLE_{0.05}$	0.004	0.008	0.945	0.672	0.017	0.000
$TLE_{0.10}$	0.001	0.029	0.980	0.005	0.008	0.885
MEM-bisquare	0.234	0.007	0.590	0.734	0.008	0.000
Mixregt	0.001	0.085	0.990	0.092	0.060	0.820

Table 4.8 ***MeSE (MSE) of Point Estimates for Unequal Variance Case with Small*** $|\gamma|$

	π		β		σ	
	5% outliers					
	MSE	MeSE	MSE	MeSE	MSE	MeSE
Hard	0.003	0.001	0.146	0.104	0.038	0.020
$Hard_{oracle}$	0.003	0.001	0.125	0.112	0.022	0.011
SCAD	0.114	0.032	5.877	3.617	1.797	1.726
$SCAD_{oracle}$	0.003	0.001	0.167	0.136	0.022	0.013
Soft	0.123	0.037	6.296	3.819	1.954	1.814
$Soft_{oracle}$	0.003	0.002	0.451	0.425	0.023	0.015
$TLE_{0.05}$	0.004	0.002	0.129	0.111	0.031	0.020
$TLE_{0.10}$	0.017	0.003	0.863	0.145	0.237	0.176
MEM-bisquare	0.183	0.005	9.725	0.193	0.443	0.123
Mixregt	0.007	0.003	0.210	0.178	0.700	0.711
MLE	0.443	0.583	16.67	18.66	5.714	2.926

Table 4.8 ***MeSE (MSE) of Point Estimates for Unequal Variance Case with Small*** $|\gamma|$ **(Continued table)**

	π		β		σ	
	10% outliers					
	MSE	MeSE	MSE	MeSE	MSE	MeSE
Hard	0.086	0.019	7.360	6.213	1.743	1.764
Hard_{oracle}	0.005	0.003	0.300	0.112	0.265	0.043
SCAD	0.150	0.077	10.98	8.265	3.037	3.005
SCAD_{oracle}	0.007	0.002	3.412	4.103	1.090	1.208
Soft	0.150	0.077	10.97	8.264	3.043	3.011
Soft_{oracle}	0.062	0.025	7.040	6.000	1.891	1.767
$\text{TLE}_{0.05}$	0.437	0.487	25.00	23.95	1.555	1.552
$\text{TLE}_{0.10}$	0.004	0.002	0.145	0.111	0.034	0.027
MEM-bisquare	0.429	0.477	22.85	22.22	2.362	2.290
Mixregt	0.074	0.004	4.298	0.303	0.513	0.444
MLE	0.310	0.361	16.44	17.62	3.480	3.613

In summary, $\text{TLE}_{0.10}$ has good results in terms of outliers detection in all cases but has larger MSE for 5% outliers case. $\text{TLE}_{0.05}$ fails to work in the case of 10% outliers due to the small α setting (less than the proportion of outliers). Hard has comparable performance to the oracle TLE and Hard_{oracle} in terms of both outlier detection and MeSE in 5% outliers case with either large or small $|\gamma|$ and in 10% outliers case with large $|\gamma|$. With small $|\gamma|$ and 10% outliers in the data, Hard_{oracle} has better performance than Hard. SCAD performs as well as Hard and SCAD_{oracle} with large $|\gamma|$. But SCAD_{oracle} performs much better than SCAD with small $|\gamma|$. Therefore, a better method to choose the tuning parameter for SCAD and HARD might improve their performance in some cases. Like MLE,

Soft is sensitive to high leverage outliers, which has also been noticed by She and Owen (2011).

4.4 Tone Perception Data Analysis

We apply the proposed robust procedure to tone perception data (Cohen, 1984). In the tone perception experiment of Cohen (1984), a pure fundamental tone with electronically generated overtones added was played to a trained musician. The experiment recorded 150 trials from the same musician. The overtones were determined by a stretching ratio, which is the ratio between adjusted tone and the fundamental tone. The purpose of this experiment was to see how this tuning ratio affects the perception of the tone and to determine if either of two musical perception theories was reasonable.

We compare our proposed Hard and traditional MLE after adding ten identical outliers (1.5, 5) into the original data set. Figure 4.1 shows the scatter plot of the data with the estimated regression lines generated by the traditional MLE (dashed lines) and the proposed Hard (solid line) for the data augmented by the outliers (stars). Based on Figure 4.1, the MLE mistakenly assigns the outliers to one component and the rest of the data to another component. In contrast, the proposed method using Hard penalty is not influenced by the added outliers and fits the two regression lines to the two correctly identified components. Using SCAD penalty leads to very similar results.

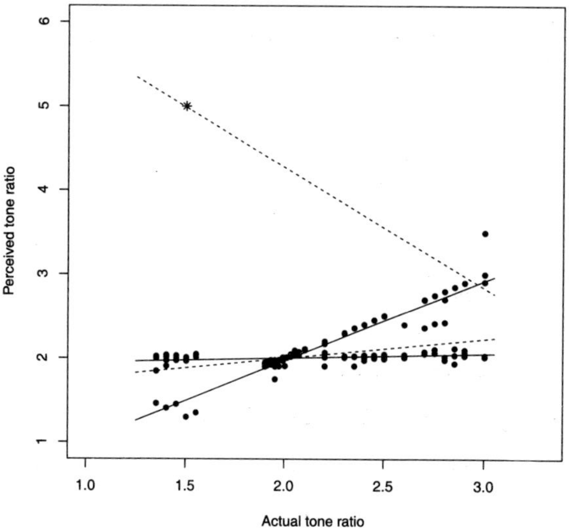

Figure 4.1 ***The scatter plot of the tone perception data and the fitted mixture regression lines with added ten identical outliers (1.5, 5) (denoted by stars at the upper left corner)***

Note: The predictor is actual tone ratio and the response is the perceived tone ratio by a trained musician. The solid lines represent the fit by the proposed Hard and the dashed lines represent the fit by the traditional MLE.

4.5 Discussion

In this chapter, we have proposed a robust mixture regression estimation procedure using mean shift model. The new model focuses on outlier detection directly and can also provide a robust model parameter estimate. Based on our simulation results, the proposed RM^2 using the hard penalty (Hard) and data adaptive chosen tuning parameter has overall comparable performance to $Hard_{oracle}$ and the oracle TLE.

In addition, note that $Hard_{oracle}$ and $SCAD_{oracle}$ have better performance than HARD and SCAD in some cases, especially when $|\gamma|$ is small. Therefore, we can further improve the performance of SCAD and HARD if having a better method to choose the tuning parameter. This requires further research. The proposed approach can be further extended to conduct simultaneous variable selection and outlier detection in mixture regression. Computationally, variable selection can be done by penalized estimation of the regression coefficients, which may then be included or embedded to the developed EM algorithm.

Appendix

Proof of Equation (3-8)

The estimate of $\boldsymbol{\gamma}$ is updated by maximizing

$$\sum_{i=1}^{n}\sum_{j=1}^{m} p_{ij}^{(k+1)} \log \phi(y_i - \gamma_i; \mu_j, \sigma^2) - \sum_{i=1}^{n} \frac{1}{w} P_\lambda(|\gamma_i|)$$

The problem is separable in each $\boldsymbol{\gamma}_i$. Thus each $\boldsymbol{\gamma}_i$ can be updated by minimizing

$$-\sum_{j=1}^{m} p_{ij}^{(k+1)} \log \phi(y_i - \gamma_i; \mu_j, \sigma^2) + \frac{1}{w} P_\lambda(|\gamma_i|)$$

Note that

$$\begin{aligned}\log \phi\left(y_i - \gamma_i; \mu_j, \sigma^2\right) &= \log \left[\left(\sigma^2\right)^{-\frac{1}{2}} \cdot \exp\left\{-\frac{\left(y_i - \gamma_i - \mu_j\right)^2}{2\sigma^2}\right\}\right] + \text{const} \\ &= -\frac{1}{2}\log\left(\sigma^2\right) - \frac{\left(y_i - \gamma_i - \mu_j\right)^2}{2\sigma^2} + \text{const}\end{aligned}$$

Thus, the solution of γ has the following form,

$$\hat{\gamma}_i = \arg\min_{\gamma_i} \sum_{j=1}^{m} p_{ij} \left\{\frac{1}{2}\log\left(\sigma^2\right) + \frac{\left(y_i - \gamma_i - \mu_j\right)^2}{2\sigma^2}\right\} + \frac{1}{w} P_\lambda\left(|\gamma_i|\right)$$

Since $\frac{1}{2}\sum_{j=1}^{m} p_{ij}\log(\sigma^2)$ does not depend on γ, we can ignore this term.

The second term is

$$\sum_{j=1}^{m} p_{ij}\frac{(y_i-\gamma_i-\mu_j)^2}{2\sigma^2}=\frac{1}{2\sigma^2}\sum_{j=1}^{m} p_{ij}\left\{\gamma_i^2-2(y_i-\mu_j)\gamma_i+(y_i-\mu_j)^2\right\}$$

$$=\frac{1}{2\sigma^2}\left[\left\{\gamma_i-\frac{\sum_{j=1}^{m} p_{ij}(y_i-\mu_j)}{\sum_{j=1}^{m} p_{ij}}\right\}^2+\text{const}\right]$$

$$=\frac{1}{2\sigma^2}\left[\left\{\gamma_i-\sum_{j=1}^{m} p_{ij}(y_i-\mu_j)\right\}^2+\text{const}\right]$$

where $\sum_{j=1}^{m} p_{ij}=1$. It follows that

$$\hat{\gamma}_i=\arg\min_{\gamma_i}\frac{1}{2\sigma^2}\left[\left\{\gamma_i-\sum_{j=1}^{m} p_{ij}(y_i-\mu_j)\right\}^2\right]+\frac{1}{w}P_\lambda(|\gamma_i|)$$

Proof of Equation (3-17)

The parameter γ is updated by maximizing

$$\sum_{i=1}^{n}\sum_{j=1}^{m} p_{ij}^{(k+1)}\log\phi(y_i-\gamma_i\sigma_j;\mu_j,\sigma_j^2)-\sum_{i=1}^{n} P_\lambda(|\gamma_i|)$$

Again, the problem is separable in each γ_i, and the estimate of each γ_i is obtained by minimizing

$$-\sum_{j=1}^{m} p_{ij}^{(k+1)}\log\phi(y_i-\gamma_i\sigma_j;\mu_j,\sigma_j^2)+P_\lambda(|\gamma_i|)$$

After some algebra, the solution of γ_i has the following form

$$\hat{\gamma}_i = \arg\min_{\gamma_i} \sum_{j=1}^{m} p_{ij} \left\{ \frac{1}{2} \log\left(\sigma_j^2\right) + \frac{\left(y_i - \gamma_i \sigma_j - \mu_j\right)^2}{2\sigma_j^2} \right\} + P_\lambda\left(|\gamma_i|\right)$$

We have that

$$\begin{aligned}\sum_{j=1}^{m} p_{ij} \frac{(y_i - \gamma_i \sigma_j - \mu_j)^2}{2\sigma_j^2} &= \sum_{j=1}^{m} \frac{p_{ij}}{2\sigma_j^2} \left\{ \gamma_i^2 \sigma_j^2 - 2\left(y_i - \mu_j\right) \gamma_i \sigma_j + \left(y_i - \mu_j\right)^2 \right\} \\ &= \frac{1}{2} \left[\left\{ \gamma_i - \frac{\sum_{j=1}^{m} \frac{p_{ij}}{\sigma_j}(y_i - \mu_j)}{\sum_{j=1}^{m} p_{ij}} \right\}^2 + \text{const} \right] \\ &= \frac{1}{2} \left[\left\{ \gamma_i - \sum_{j=1}^{m} \frac{p_{ij}}{\sigma_j}(y_i - \mu_j) \right\}^2 + \text{const} \right]\end{aligned}$$

where $\sum_{j=1}^{m} p_{ij} = 1$. It follows that

$$\hat{\gamma}_i = \arg\min_{\gamma_i} \frac{1}{2} \left[\left\{ \gamma_i - \sum_{j=1}^{m} \frac{p_{ij}}{\sigma_j}(y_i - \mu_j) \right\}^2 \right] + P_\lambda\left(|\gamma_i|\right)$$

Proof of SCAD thresholding rule in Proposition 3-1

The penalized least squares has the following form:

$$\frac{1}{2}(\gamma - \xi)^2 + \frac{\sigma^2}{\hat{\sigma}^2} P_\lambda(\gamma)$$

where

$$\xi = \frac{\sum_{j=1}^{m} p_{ij}(y_i - \mu_j)}{\sum_{j=1}^{m} p_{ij}}$$

Note that for simplicity, we have omitted the subscripts in γ_i and ξ_i.

Consider the first derivative of the above penalized least squares with respect to γ

$$\frac{\partial\left\{\frac{1}{2}(\gamma-\xi)^2+\frac{\sigma^2}{\hat{\sigma}^2}P_\lambda(\gamma)\right\}}{\partial\gamma}=\gamma-\xi+sgn(\gamma)\frac{\sigma^2}{\hat{\sigma}^2}P'_\lambda(\gamma)$$

where

$$P'_\lambda(\gamma)=\begin{cases}\lambda & \text{if } 0<|\gamma|\leqslant\lambda\\ \frac{(a\lambda-|\gamma|)_+}{a-1} & \text{if } \lambda<|\gamma|\leqslant a\lambda\\ 0 & \text{if } |\gamma|>a\lambda\end{cases}$$

We shall check the second derivative of the above penalized least squares in three cases.

Case 1: when $0<|\gamma|\leqslant\lambda$

$$\frac{\partial\left\{(\gamma-\xi)+sgn(\gamma)\frac{\sigma^2}{\hat{\sigma}^2}P'_\lambda(\gamma)\right\}}{\partial\gamma}=\frac{\partial\left(\gamma-\xi+sgn(\gamma)\frac{\sigma^2\lambda}{\hat{\sigma}^2}\right)}{\partial\gamma}=1>0$$

Solving the equation $\gamma-\xi+sgn(\gamma)\dfrac{\sigma^2\lambda}{\hat{\sigma}^2}=0$, we have $\hat{\gamma}=\xi-\dfrac{\sigma^2\lambda}{\hat{\sigma}^2}$ and $\hat{\gamma}=-(-\xi-\dfrac{\sigma^2\lambda}{\hat{\sigma}^2})=\xi+\dfrac{\sigma^2\lambda}{\hat{\sigma}^2}$.

Case 2: when $\lambda<|\gamma|\leqslant a\lambda$

$$\frac{\partial\left\{(\gamma-\xi)+sgn(\gamma)\frac{\sigma^2}{\hat{\sigma}^2}P'_\lambda(\gamma)\right\}}{\partial\gamma}=\frac{\partial\left\{\gamma-\xi+sgn(\gamma)\frac{\sigma^2(a\lambda-|\gamma|)}{\hat{\sigma}^2(a-1)}\right\}}{\partial\gamma}=1-\frac{\sigma^2}{\hat{\sigma}^2(a-1)}$$

If $\dfrac{\sigma^2}{\hat{\sigma}^2}<\alpha-1$, then the second derivative is positive. Solving the equation $\gamma-\xi+sgn(\gamma)\dfrac{\sigma^2(a\lambda-\gamma)}{\hat{\sigma}^2(a-1)}=0$, we have $\hat{\gamma}=\dfrac{\frac{\hat{\sigma}^2}{\sigma^2}(a-1)\xi-a\lambda}{\frac{\hat{\sigma}^2}{\sigma^2}(a-1)-1}$ and $\hat{\gamma}=$

$$-\left\{\frac{\frac{\hat{\sigma}^2}{\sigma^2}(a-1)(-\xi)-a\lambda}{\frac{\hat{\sigma}^2}{\sigma^2}(a-1)-1}\right\}=\frac{\frac{\hat{\sigma}^2}{\sigma^2}(a-1)\xi+a\lambda}{\frac{\hat{\sigma}^2}{\sigma^2}(a-1)-1}$$

If $\frac{\sigma^2}{\hat{\sigma}^2}>a-1$, then the second derivative is negative and the solution of the equation $\gamma-\xi+sgn(\gamma)\frac{\sigma^2(a\lambda-\gamma)}{\hat{\sigma}^2(a-1)}=0$ is not a minimizer of the penalized least squares.

Case 3: when $|\gamma|>a\lambda$,

$$\frac{\partial\left\{(\gamma-\xi)+sgn(\gamma)\frac{\sigma^2}{\hat{\sigma}^2}P_\lambda'(\gamma)\right\}}{\partial\gamma}=\frac{\partial(\gamma-\xi)}{\partial\gamma}=1>0$$

Solving the equation $\gamma-\xi=0$, we have $\hat{\gamma}=\xi$.

From the above three cases, we can see that the γ solutions depend on the values of $\frac{\sigma^2}{\hat{\sigma}^2}$ and ξ. Next, we must verify γ solutions in the following scenarios:

When $\sigma^2/\hat{\sigma}^2<a-1$

Note: For a positive λ, $\frac{\sigma^2}{\hat{\sigma}^2}<a-1$ is equivalent to $\lambda+\frac{\sigma^2\lambda}{\hat{\sigma}^2}<a\lambda$. Since the penalized least squares is symmetric and $\Theta(-\xi;\lambda)=-\Theta(\xi;\lambda)$, we have $\hat{\gamma}=\Theta(-\xi;\lambda)=-\Theta(\xi;\lambda)$. Here we only discuss positive ξ.

1. When $\xi>a\lambda$, γ satisfies Case 3, then we have $\hat{\gamma}=\xi$.

2. When $\lambda+\frac{\sigma^2\lambda}{\hat{\sigma}^2}<\xi\leqslant a\lambda$, γ satisfies Case 2, then we have $\hat{\gamma}=\frac{\frac{\hat{\sigma}^2}{\sigma^2}(a-1)\xi-a\lambda}{\frac{\hat{\sigma}^2}{\sigma^2}(a-1)-1}$.

3. When $\frac{\sigma^2\lambda}{\hat{\sigma}^2}<\xi\leqslant\lambda+\frac{\sigma^2\lambda}{\hat{\sigma}^2}$, γ satisfies Case 1, then we have $\hat{\gamma}=\xi-\frac{\sigma^2\lambda}{\hat{\sigma}^2}$.

4. For $0\leqslant\xi\leqslant\frac{\sigma^2\lambda}{\hat{\sigma}^2}$, γ satisfies Case 1. If $\gamma\geqslant 0$, the first derivative of penalized least squares, $\gamma-\xi+\frac{\sigma^2\lambda}{\hat{\sigma}^2}$, is monotone increasing, so $\hat{\gamma}=0$; similarly, if $\gamma\leqslant 0$, the first derivative of penalized least squares is monotone decreasing, so $\hat{\gamma}=0$.

In summary, we have

$$\hat{\gamma}=\begin{cases} sgn(\xi)\left(|\xi|-\frac{\sigma^2\lambda}{\hat{\sigma}^2}\right)_+, & \text{if } |\xi|\leqslant\lambda+\frac{\sigma^2\lambda}{\hat{\sigma}^2} \\ \frac{\frac{\hat{\sigma}^2}{\sigma^2}(a-1)\xi-sgn(\xi)a\lambda}{\frac{\hat{\sigma}^2}{\sigma^2}(a-1)-1}, & \text{if } \lambda+\frac{\sigma^2\lambda}{\hat{\sigma}^2}<|\xi|\leqslant a\lambda \\ \xi, & \text{if } |\xi|>a\lambda \end{cases}$$

When $a-1\leqslant\sigma^2/\hat{\sigma}^2\leqslant a+1$

Note: For a positive λ, $\frac{\sigma^2}{\hat{\sigma}^2}\geqslant a-1$ is equivalent to $\lambda+\frac{\sigma^2\lambda}{\hat{\sigma}^2}\geqslant a\lambda$. We consider the following subcases:

1. When $|\xi|\leqslant a\lambda$, based on the result summary when $\sigma^2/\hat{\sigma}^2<a-1$,

$$\hat{\gamma}=sgn(\xi)\left(|\xi|-\frac{\sigma^2\lambda}{\hat{\sigma}^2}\right)_+$$

2. When $a\lambda\leqslant|\xi|\leqslant\lambda+\frac{\sigma^2\lambda}{\hat{\sigma}^2}$, for $\hat{\gamma}=sgn(\xi)\left(|\xi|-\frac{\sigma^2\lambda}{\hat{\sigma}^2}\right)_+$, the objective function becomes

$$f_1=\frac{1}{2}(\hat{\gamma}-\xi)^2+\frac{\sigma^2}{\hat{\sigma}^2}\lambda|\hat{\gamma}|$$

and for $\hat{\gamma}_2=\xi$, the objective function becomes

$$f_2=\frac{\sigma^2(a+1)\lambda^2}{2\hat{\sigma}^2}$$

Define $d=f_1-f_2$. If $d>0$, then $\hat{\gamma}=\xi$. If $d<0$, then $\hat{\gamma}=sgn(\xi)\left(|\xi|-\frac{\sigma^2\lambda}{\hat{\sigma}^2}\right)_+$.

(i) When $\xi>\frac{\sigma^2\lambda}{\hat{\sigma}^2}$, $\hat{\gamma}_1=\xi-\frac{\sigma^2\lambda}{\hat{\sigma}^2}$ and

$$f_1=\frac{1}{2}\frac{\sigma^2\lambda^2}{\hat{\sigma}^2}+\frac{\sigma^2}{\hat{\sigma}^2}\lambda(\xi-\frac{\sigma^2\lambda}{\hat{\sigma}^2})$$

Then

$$d=f_1-f_2=\frac{\sigma^2\lambda^2}{2\hat{\sigma}^2}\left(\frac{2\xi}{\lambda}-a-1-\frac{\sigma^2}{\hat{\sigma}^2}\right)$$

When $\xi>\frac{a+1+\frac{\sigma^2}{\hat{\sigma}^2}}{2}\lambda$, $d>0$, so $\hat{\gamma}=\xi$. When $\xi<\frac{a+1+\frac{\sigma^2}{\hat{\sigma}^2}}{2}\lambda$, $d<0$, so $\hat{\gamma}=\xi\frac{\sigma^2\lambda}{\sigma^2}$.

Note that since $\frac{\sigma^2}{\hat{\sigma}^2}>a-1$, $\frac{a+1+\frac{\sigma^2}{\hat{\sigma}^2}}{2}\lambda>a\lambda$.

Note that in order to result in the soft thresholding rule $\hat{\gamma}=sgn(\xi)\left(|\xi|-\frac{\sigma^2\lambda}{\hat{\sigma}^2}\right)$, we need $\frac{\sigma^2\lambda}{\hat{\sigma}^2}\leqslant\frac{a+1+\frac{\sigma^2}{\hat{\sigma}^2}}{2}\lambda$ i. e., $\left[-\frac{\sigma^2\lambda}{\hat{\sigma}^2},\frac{\sigma^2\lambda}{\hat{\sigma}^2}\right]$ is contained within $\left[-\frac{a+1+\frac{\sigma^2}{\hat{\sigma}^2}}{2}\lambda,\frac{a+1+\frac{\sigma^2}{\hat{\sigma}^2}}{2}\lambda\right]$. Accordingly, $\frac{\sigma^2\lambda}{\hat{\sigma}^2}\leqslant\frac{a+1+\frac{\sigma^2}{\hat{\sigma}^2}}{2}\lambda$ indicates $\frac{\sigma^2}{\hat{\sigma}^2}\leqslant a+1$.

(ii) When $0\leqslant\xi\leqslant\frac{\sigma^2\lambda}{\hat{\sigma}^2}$, $\hat{\gamma}_1=0$, and $f_1=\frac{\xi^2}{2}$.

$$d = f_1 - f_2 = \frac{\xi^2}{2} - \frac{\sigma^2(a+1)\lambda^2}{2\hat{\sigma}^2}$$
$$< \frac{\sigma^2\lambda^2}{2\hat{\sigma}^2}\left\{\frac{\sigma^2}{\hat{\sigma}^2} - (a+1)\right\}$$

since $\frac{\sigma^2}{\hat{\sigma}^2} \leqslant a+1$, $d<0$ $\hat{\gamma}=0$.

3. When $|\xi|>\lambda+\frac{\sigma^2\lambda}{\hat{\sigma}^2}$, based on the result summary when $\sigma^2/\hat{\sigma}^2 < a-1$, $\hat{\gamma}=\xi$. By summarizing the above three subcases and symmetry property, we have

$$\hat{\gamma} = \begin{cases} sgn(\xi)\left(|\xi| - \frac{\sigma^2\lambda}{\hat{\sigma}^2}\right)_+, & \text{if } |\xi| \leqslant \frac{a+1+\frac{\sigma^2}{\hat{\sigma}^2}}{2}\lambda \\ \xi, & \text{if } |\xi| > \frac{a+1+\frac{\sigma^2}{\hat{\sigma}^2}}{2}\lambda \end{cases}$$

When $\sigma^2/\hat{\sigma}^2>a+1$

For $\frac{\sigma^2}{\hat{\sigma}^2}>a+1$, we have $\frac{\sigma^2\lambda}{\hat{\sigma}^2}>\frac{a+1+\frac{\sigma^2}{\hat{\sigma}^2}}{2}\lambda$. Consider the following two subcases:

1. When $\xi>\frac{\sigma^2\lambda}{\hat{\sigma}^2}>\frac{a+1+\frac{\sigma^2}{\hat{\sigma}^2}}{2}\lambda$, $\hat{\gamma}=\xi$.

2. When $0\leqslant\xi\leqslant\frac{\sigma^2\lambda}{\hat{\sigma}^2}$, $\hat{\gamma}_1=0$ and

$$d = f_1 - f_2 = \frac{\xi^2}{2} - \frac{\sigma^2(a+1)\lambda^2}{2\hat{\sigma}^2}.$$

If $|\xi|<\sqrt{\frac{\sigma^2(a+1)}{\hat{\sigma}^2}}\lambda$, $d<0$, then $\hat{\gamma}=0$; If $|\xi|>\sqrt{\frac{\sigma^2(a+1)}{\hat{\sigma}^2}}\lambda$, $d>0$, then $\hat{\gamma}=\xi$.

By summarizing the above two subcases and symmetry property, we have

$$\hat{\gamma}=\begin{cases}0, & \text{if } |\xi| \leqslant \sqrt{\frac{\sigma^2(a+1)}{\hat{\sigma}^2}}\lambda \\ \xi, & \text{if } |\xi| > \sqrt{\frac{\sigma^2(a+1)}{\hat{\sigma}^2}}\lambda\end{cases}$$

Proof of Equation (4-6)

The estimate of $\boldsymbol{\gamma}$ is updated based on updated p_{ij}, $\boldsymbol{\beta}$ and $\boldsymbol{\sigma}_j$ by maximizing

$$\sum_{i=1}^{n}\sum_{j=1}^{m} p_{ij}^{(k+1)} \log \phi(y_i - \boldsymbol{\gamma}_i\sigma_j - \mathbf{x}_i^T\boldsymbol{\beta}_j; 0, \sigma_j^2) - \sum_{i=1}^{n} P_\lambda(|\gamma_i|)$$

To do this, each $\boldsymbol{\gamma}_i$ is separately updated by maximizing:

$$\sum_{j=1}^{m} p_{ij}^{(k+1)} \log \phi(y_i - \gamma_i\sigma_j - \mathbf{x}_i^T\boldsymbol{\beta}_j; 0, \sigma_j^2) - P_\lambda(|\gamma_i|)$$

Equivalently, the estimate of $\boldsymbol{\gamma}_i$ is updated by minimizing

$$-\sum_{j=1}^{m} p_{ij}^{(k+1)} \log \phi(y_i - \gamma_i\sigma_j - \mathbf{x}_i^T\boldsymbol{\beta}_j; 0, \sigma_j^2) + P_\lambda(|\gamma_i|)$$

Note that

$$\begin{aligned}&\log \phi\left(y_i - \gamma_i\sigma_j - \mathbf{x}_i^T\boldsymbol{\beta}_j; 0, \sigma_j^2\right)\\ &= \log\left[\left(\sigma_j^2\right)^{-\frac{1}{2}} \exp\left\{-\frac{\left(y_i - \gamma_i\sigma_j - \mathbf{x}_i^T\boldsymbol{\beta}_j\right)^2}{2\sigma_j^2}\right\}\right] + const\\ &= -\frac{1}{2}\log\left(\sigma_j^2\right) - \frac{\left(y_i - \gamma_i\sigma_j - \mathbf{x}_i^T\boldsymbol{\beta}_j\right)^2}{2\sigma_j^2} + const\end{aligned}$$

Thus, the solutions of $\boldsymbol{\gamma}$ have the following form:

$$\gamma_i = \arg\min \frac{1}{2}\sum_{j=1}^{m} p_{ij}\log\left(\sigma_j^2\right) + \sum_{j=1}^{m} p_{ij}\left\{\frac{\left(y_i - \gamma_i\sigma_j - \mathbf{x}_i^T\boldsymbol{\beta}_j\right)^2}{2\sigma_j^2}\right\} + P_\lambda\left(|\gamma_i|\right)$$

Since $\frac{1}{2}\sum_{j=1}^{m} p_{ij}\log\left(\sigma_j^2\right)$ does not depend on $\boldsymbol{\gamma}$, we can ignore this term.

The second term:

$$\begin{aligned}
&\sum_{j=1}^{m} p_{ij}\frac{(y_i-\gamma_i\sigma_j-\mathbf{x}_i^T\boldsymbol{\beta}_j)^2}{2\sigma_j^2}\\
&=\sum_{j=1}^{m}\frac{p_{ij}}{2\sigma_j^2}\left\{\gamma_i^2\sigma_j^2-2\left(y_i-\mathbf{x}_i^T\boldsymbol{\beta}_j\right)\gamma_i\sigma_j+\left(y_i-\mathbf{x}_i^T\boldsymbol{\beta}_j\right)^2\right\}\\
&=\frac{1}{2}\left[\left\{\gamma_i-\frac{\sum_{j=1}^{m}\frac{p_{ij}}{\sigma_j}(y_i-\mathbf{x}_i^T\boldsymbol{\beta}_j)}{\sum_{j=1}^{m}p_{ij}}\right\}^2+constant\right]\\
&=\frac{1}{2}\left[\left\{\gamma_i-\sum_{j=1}^{m}\frac{p_{ij}}{\sigma_j}(y_i-\mathbf{x}_i^T\boldsymbol{\beta}_j)\right\}^2+constant\right]
\end{aligned}$$

where $\sum_{j=1}^{m} p_{ij}=1$.

Therefore,

$$\gamma_i=\arg\min\frac{1}{2}\left[\left\{\gamma_i-\sum_{j=1}^{m}\frac{p_{ij}}{\sigma_j}(y_i-\mathbf{x}_i^T\boldsymbol{\beta}_j)\right\}^2\right]+P_\lambda\left(|\gamma_i|\right)$$

References

[1] Andrews D. F. and Mallows C. L. Scale mixtures of normal distributions [J]. *Journal of the Royal Statistical Society, Series B*, 1974 (36): 99–102.

[2] Antoniadis A. Wavelets in Statistics: A Review (with discussion) [J]. *Journal of the Italian Statistical Association*, 1997 (6): 97–144.

[3] Bai X., Yao W., and Boyer J. E. Robust fitting of mixture regression models [J]. *Computational Statistics and Data Analysis*, 2012 (56): 2347–2359.

[4] Bashir S. and Carter E. Robust mixture of linear regression models [J]. *Communications in Statistics–Theory and Methods*, 2012 (41): 3371–3388.

[5] Basso R. M., Lachos V. H., Cabral C. R. B. and Ghosh P. Robust mixture modeling based on scale mixtures of skew-normal distributions [J]. *Computational Statistics and Data Analysis*, 2010 (54): 2926–2941.

[6] Böhning D. Computer–Assisted Analysis of Mixtures and Applications [M]. Boca Raton, FL: Chapman and Hall/CRC, 1999.

[7] Carroll R. J. and Welsch A. H. A Note on Asymmetry and Robustness in Linear Regression [J]. *Journal of American Statisitcal Association*, 1988 (4): 285–287.

[8] Celeux G., Hurn M., and Robert C. P. Computational and inferential

difficulties with mixture posterior distributions [J]. *Journal of the American Statistical Association* , 2000 (95): 957-970.

[9] Chen J., Tan X., and Zhang R. Inference for normal mixture in mean and variance [J]. *Statistica Sincia*, 2008 (18): 443-465.

[10] Chen Y., Wang Q., and Yao W. Adaptive Estimation for Varying Coefficient Models [J]. *Journal of Multivariate Analysis*, 2015 (137): 17-31.

[11] Coakley C. W. and Hettmansperger T. P. A Bounded Influence, High Breakdown, Efficient Regression Estimator [J]. *Journal of American Statistical Association* , 1993 (88): 872-880.

[12] Cohen E. Some effects of inharmonic partials on interval perception [J]. *Music Perception* , 1984 (1): 323-349.

[13] Crawford S. L., Degroot M. H., Kadane J. B., and Small M. J. Modeling lake-chemistry distributions-approximate Bayesian methods for estimating a finite-mixture model [J]. *Technometrics*, 1992 (34): 441-453.

[14] Crawford S. L. An application of the Laplace method to finite mixture distributions [J]. *Journal of the American Statistical Association*, 1994 (89): 259-267.

[15] Croux C., Rousseeuw P. J., and Hössjer O. Generalized S-estimators [J]. *Journal of American Statistical Association*, 1994 (89): 1271-1281.

[16] Dempster A. P., Laird N. M., and Rubin D. B. Maximum likelihood from incomplete data via the EM algorithm (with discussion) [J]. *Journal of the Royal Statistical Society*, *Series B*, 1977 (39): 1-38.

[17] Doğru F. Z. and Arslan O. Robust mixture regression modeling using the least trimmed squares (LTS)-estimation method [J]. *Communications in Sta-*

tistics –Simulation and Computation Forthcoming, 2017.

[18] Donoho D. L. and Huber P. J. The Notation of Break–down Point [J]. *in A Festschrift for E. L. Lehmann, Wadsworth*, 1983.

[19] Donoho D. L. and Johnstone I. M. Ideal Spatial Adaptation by Wavelet shrinkage [J]. *Biometrika*, 1994a (81): 425–455.

[20] Everitt B. S. and Hand D. J. Finite Mixture Distributions [M]. London: Chapman and Hall, 1981.

[21] Fan J. and Li R. Variable selection via nonconcave penalized likelihood and its oracle properties [J]. *Journal of the American Statistical Association*, 2001 (96): 1348–1360.

[22] Frühwirth–Schnatter S. *Finite Mixture and Markov Switching Models* [M]. Springer, 2006.

[23] Fujisawa H. and Eguchi S. Robust estimation in the normal mixture model [J]. *Journal of Statistical Planning and Inference*, 2015: 1–23.

[24] García–Escudero L. A., Gordaliza A., Mayo–Iscara A., and San Martín R. Robust clusterwise linear regression through trimming [J]. *Computational Statistics and Data Analysis*, 2010 (54): 3057–3069.

[25] García–Escudero L. A., Gordaliza A., Greselin F., Ingrassia S., and Mayo–Iscara A. Robust estimation of mixture regression with random covariates, via trimming and constraints [J]. *Statistics and Computing*, 2017 (27): 377–402.

[26] Gervini D. and Yohai V. J. A Class of Robust and Fully Efficient Regression Estimators [J]. *The Annals of Statistics*, 2002 (30): 583–616.

[27] Goldfeld S. M. and Quandt R. E. A Markov model for switching regression [J]. *Journal of Econometrics*, 1973 (1): 3–15.

[28] Hadi A. S., Simonoff J. S. Procedures for the identification of multiple outliers in linear models [J]. *Journal of the American Statistical Association*, 1993 (88): 1264–1272.

[29] Handschin E., Kohlas J., Fiechter A., and Schweppe F. Bad Data Analysis for Power System State Estimation [J]. *IEEE Transactions on Power Apparatus and Systems*, 1975 (2): 329–337.

[30] Hawkins D. M., Bradu D., Kass G. V. Location of several outliers in multiple regression data using elemental sets [J]. *Technometrics*, 1984 (26): 197–208.

[31] Hawkins D. M. and Olive D. J. Inconsistency of resampling algorithms for high break-down regression estimators and a new algorithm [J]. *Journal of the American Statistical Association*, 2002 (97): 136–148.

[32] Hathaway R. J. A constrained formulation of maximum–likelihood estimation for normal mixture distributions [J]. *Annals of Statistics*, 1985 (13): 795–800.

[33] Hathaway R. J. A constrained EM algorithm for univariate mixtures [J]. *Journal of Statistical Computation and Simulation*, 1986 (23): 211–230.

[34] Hennig C. Identifiability of models for clusterwise linear regression [J]. *Journal of Classification*, 2000 (17): 273–296.

[35] Hennig C. Fixed point clusters for linear regression: computation and comparison [J]. *Journal of Classification*, 2002 (19): 249–276.

[36] Hennig C. Clusters, outliers, and regression: Fixed point clusters [J]. *Journal of Multivariate Analysis*, 2003 (86): 183–212.

[37] Hu H., Yao W. and Wu Y. The robust EM–type algorithms for log–

concave mixtures of regression models [J]. *Computational Statistics and Data Analysis*, 2017 (111): 14-26.

[38] Huber P. J. Robust Version of a Location Parameter [J]. *Annals of Mathematical Statistics*, 1964 (36): 1753-1758.

[39] Huber P. J. Robust Statistics [M]. *New York: John Wiley and Sons*, 1981.

[40] Jackel L. A. Estimating Regression Coefficients by Minimizing the Dispersion of the Residuals [J]. *Annals of Mathematical Statistics*, 1972 (5): 1449-1458.

[41] Jiang W. and Tanner M. A. Hierarchical mixtures-of-experts for exponential family regression models: Approximation and maximum likelihood estimation [J]. *The Annals of Statistics*, 1999 (27): 987-1011.

[42] Kemp G. C. R. and Santos Silva J. M. C. Regression towards the mode [J]. *Journal of Economics*, 2012 (170): 92-101.

[43] Lee M. J. Mode Regression [J]. *Journal of Econometrics*, 1989 (42): 337-349.

[44] Lee M. J. Quadratic Mode Regression [J]. *Journal of Econometrics*, 1993 (57): 1-19.

[45] Lee Y., MacEachern S. N., Jung Y. Regularization of Case-Specific Parameters for Robustness and Efficiency [J]. *Statistical Science*, 2012 (27): 350-372.

[46] Linton O. B. and Xiao Z. A nonparametric regression estimator that adapts to error distribution of unknown form [J]. *Econometric Theory*, 2007 (23): 371-413.

[47] Jiang W. and Tanner M. A. Hierarchical mixtures-of-experts for exponential family regression models: Approximation and maximum likelihood estimation [J]. *The Annals of Statistics*, 1999 (27): 987-1011.

[48] Lindsay B. G. Mixture Models: Theory, Geometry and Applications, *NSF-CBMS Regional Conference Series in Probability and Statistics* Vol. 5. Institute of mathematical Statistics and the American Statisitcal Association, Alexandra, VA. 1995.

[49] Mallows C. L. On Some Topics in Robustness. Unpublished memorandum, Bell Tel. Laboratories, Murray Hill. 1975.

[50] Markatou M. Mixture models, robustness, and the weighted likelihood methodology [J]. *Biometrics*, 2000 (56): 483-486.

[51] Maronna R. A., Martin R. D., and Yohai V. J. Robust Statistics: Theory and Methods [M]. New York: Wiley, 2006.

[52] McLachlan G. J. and Basford K. E. Mixture Models: Inference and Applications to Clustering [M]. New York: Marcel Dekker, 1988.

[53] McLachlan G. J. and Peel D. Finite Mixture Models [M]. New York: Wiley, 2000.

[54] Naranjo J. D., Hettmansperger T. P. Bounded Influence Rank Regression [J]. *Journal of the Royal Statistical Society B*, 1994 (56): 209-220.

[55] Neykov N., Filzmoser P., Dimova R., and Neytchev P. Robust fitting of mixtures using the trimmed likelihood estimator [J]. *Computational Statistics and Data Analysis*, 2007 (52): 299-308.

[56] Olive D. J. and Hawkins D. M. Practical high breakdown regression [EB/OL]. http://lagrange. math. siu. edu/Olive/pphbreg. pdf, 2011.

[57] Park Y., Kim D. and Kim S. Robust regression using data partitioning and Mestimation [J]. *Communications in Statistics –Simulation and Computation*, 2012 (41): 1282–1300.

[58] Peel D. and McLachlan G. J. Robust mixture modelling using the t distribution [J]. *Statistics and Computing*, 2000 (10): 339–348.

[59] Punzo A. and McNicholas P. D. Robust clustering in regression analysis via the contam–inated gaussian cluster–weighted model [J]. *Journal of Classification*, 2017 (34): 249–293.

[60] Qin Z., Scheinberg K., and Goldfarb D. Efficient block–coordinate descent algorithms for the Group Lasso [J]. *Mathematical Programming Computation*, 2013 (5): 143–169.

[61] Richardson S. and Green P. J. On Bayesian analysis of mixtures with an unknown number of components (with discussion) [J]. *Journal of The Royal Statistical Society Series* B, 1997 (59): 731–792.

[62] Rousseeuw P. J. Least Median of Squares regression. Resaerch Report No. 178, Centre for Statistics and Operations research, VUB Brussels, 1982.

[63] Rousseeuw P. J. Multivariate Estimation with High Breakdown Point. Resaerch Report No. 192, Center for Statistics and Operations research, VUB Brussels, 1983.

[64] Rousseeuw P. J. Least Median of Squares regression [J]. *Journal of American Statistical Association*, 1984 (79): 871–880.

[65] Rousseeuw P. J. and Yohai V. J. Robust Regression by Means of S–estimators [J]. *Robust and Nonlinear Time series* J. Franke, W. Härdle and R. D. Martin (eds.), Lectures Notes in Statistics, 1984 (26): 256–272, New York:

Springer.

[66] Rousseeuw P. J. and Leroy A. M. Robust Regression and Outlier Detection [M]. New York: Wiley, 1987.

[67] Rousseeuw P. J. and Croux C. Alternatives to the Median Absolute Deviation [J]. *Journal of American Statistical Association*, 1993 (94): 388-402.

[68] Rousseeuw P. J. and Van Driessen K. A fast algorithm for the minimum covariance determinant estimator [J]. *Technometrics*, 1999 (41): 212-223.

[69] She Y. Thresholding-Based Iterative Selection Procedures for Model Selection and Shrinkage [J]. *Electronic Journal of Statistics*, 2009 (3): 384-415.

[70] She Y. and Owen A. Outlier Detection Using Nonconvex Penalized Regression [J]. *Journal of the American Statistical Association*, 2011 (106): 626-639.

[71] Shen H., Yang J., and Wang S. Outlier detecting in fuzzy switching regression models [J]. *Artificial Intelligence: Methodology, Systems, and Applications Lecture Notes in Computer Science*, 2004 (3192): 208-215.

[72] Siegel A. F. Robust Regression Using Repeated Medians [J]. *Biometrika*, 1982 (69): 242-244.

[73] Song W., Yao, W., and Xing Y. Robust mixture regression model fitting by laplace distribution [J]. *Computational Statistics and Data Analysis*, 2014 (71): 128-137.

[74] Stromberg A. J., Hawkins, D. M., and Hössjer, O. The Least Trimmed Differences Regression Estimator and Alternatives [J]. *Journal of American Statistical Association*, 2000 (95): 853-864.

[75] Skrondal A. and Rabe-Hesketh, S. *Generalized Latent Variable Modeling: Multilevel, Longitudinal and Structural Equation Models* [M]. Boca Raton.

Chapman and Hall/CRC, 2004.

[76] Stephens, M. Dealing with label switching in mixture models [J]. *Journal of The Royal Statistical Society Series* B, 2000 (62): 795-809.

[77] Sun J., Kabán, A. and Garibaldi J. M. Robust mixture clustering using Pearson type Ⅶ distribution [J]. *Pattern Recognition Letters*, 2010 (31): 2447-2454.

[78] Tibshirani, R. J. Regression Shrinkage and Selection via the LASSO [J]. *Journal of The Royal Statistical Society Series* B, 1996 (58): 267-288.

[79] Tibshirani R. J. The LASSO Method for Variable Selection in the Cox Model [J]. *Statistics in Medicine*, 1996 (16): 385-395.

[80] Titterington D. M., Smith, A. F. M., and Makov, U. E. Statistical Analysis of Finite Mixture Distribution [M]. Wiley, New York, 1985.

[81] Wang Q. and Yao, W. An adaptive estimation of MAVE [J]. *Journal of Multivariate Analysis*, 2012 (104): 88-100.

[82] Wedel M. and Kamakura, W. A. *Market Segmentation: Conceptual and Methodological Foundations* [M]. 2nd edition, Norwell, MA: Kluwer Academic Publishers. Journal of Classification. Springer, New York, 2000.

[83] Wilcox. A review of some recent developments in robust regression [J]. *British Journal of Mathematical and Statistical Psychology*, 1996 (49): 253-274.

[84] Yao W. A profile likelihood method for normal mixture with unequal variance [J]. *Journal of Statistical Planning and Inference*, 2010 (140): 2089-2098.

[85] Yao W. Model based labeling for mixture models [J]. *Statistics and Computing*, 2012 (22): 337-347.

[86] Yao W. and Lindsay B. G. Bayesian mixture labeling by highest posterior density [J]. *Journal of American Statistical Association*, 2009 (104): 758-767.

[87] Yao W. and Li L. A new regression model: modal linear regression [J]. *Scandinavian Journal of Statistics*, 2014: 1-16.

[88] Yao W. and Zhao, Z. Kernel density based linear regression estimates [J]. *Communications in Statistics-Theory and Methods*, 2013 (42): 4499-4512.

[89] Yao W., Wei Y., and Yu C. Robust mixture regression using the t-distribution [J]. *Computational Statistics and Data Analysis* 2014 (71): 116-127.

[90] Yi G. Y., Tan X., and Li R. Variable selection and inference procedures for marginal analysis of longitudinal data with missing observations and covariate measurement error [J]. *The Canadian Journal of Statistics*, 2015 (43): 498-518.

[91] Yohai V. J. High Breakdown-point and High Efficiency Robust Estimates for Regression [J]. *The Annals of Statistics*, 1987 (15): 642-656.

[92] You J. A Monte Carlo Comparison of Several High Breakdown and Efficient Estimators [J]. *Computational Statistics and Data Analysis*, 1999 (30): 205-219.

[93] Yuan A. and De Gooijer J. G. Semiparametric regression with kernel error model [J]. *Scandinavian Journal of Statistics*, 2007 (34): 841-869.

[94] Zhang J. Robust mixture regression modeling with Pearson type Ⅶ distribution [D]. Kansas State University, 2013.

[95] Zeller C. B., Cabral C. R. and Lachos V. H. Robust mixture regression modeling based on scale mixtures of skew-normal distributions [J]. *Test*,

2016 (25): 375-396.

[96] Zou H. The Adaptive Lasso and Its Oracle Properties [J]. *Journal of American Statistical Association*, 2006 (101): 1418-1429.